IT
ALL
ADDS
UP

IT ALL ADDS UP

THE STORY OF PEOPLE AND MATHEMATICS

MICKAËL LAUNAY

TRANSLATED BY
STEPHEN S WILSON

WILLIAM
COLLINS

William Collins
An imprint of HarperCollinsPublishers
1 London Bridge Street
London
SE1 9GF

www.WilliamCollinsBooks.com

First published in Great Britain by William Collins in 2018

First published in France by Flammarion,
as *Le grand romans des maths* in 2016

1

Drawing on page 3 by Maurice Bourlon; Image on page 22 © IRSNB, Thierry Hubin
Map on page 108 in the public domain, courtesy of Bibliothèque nationale de France,
GE BB-565 (A7,10); Image on page 209 from Wikimedia Commons;
Image on page 222 by Stefan Zachow.

A catalogue record for this book is available from the British Library

ISBN 978-0-00-828393-3 (hardback)
ISBN 978-0-00-828394-0 (trade paperback)

Typeset in Minion Pro by Palimpsest Book Production Ltd, Falkirk, Stirlingshire
Printed and bound in Great Britain by CPI Group (UK) Ltd, Croydon CR0 4YY

MIX
Paper from
responsible sources
FSC
www.fsc.org   FSC® C007454

This book is produced from independently certified FSC™ paper
to ensure responsible forest management.

For more information visit: www.harpercollins.co.uk/green

CONTENTS

FOREWORD

'Oh, I've never been much good at maths myself!'

I'm getting a little blasé. This must be at least the tenth time I've heard that remark today.

But this lady has been here at my stall for a good fifteen minutes, standing with a group of other visitors, listening attentively while I describe various geometrical curiosities. That's how the conversation started.

'But what do you do for a living?' she asked me.

'I'm a mathematician.'

'Oh, I've never been much good at maths myself!'

'Really? But you seemed to be interested in what I was just talking about.'

'Yes . . . but that's not really maths . . . that was understandable.'

I hadn't heard that one before. Is mathematics, by definition, a discipline that can't be understood?

It's the beginning of August, in the Cours Félix Faure in La Flotte-en-Ré, France. In this small summer market, I have a pop-up – there is henna tattooing and afro braids to my right, a mobile-phone accessory stall to my left, and a display of jewels and trinkets of all kinds opposite me. I've set up my maths stand in the middle of all this. Holidaymakers stroll peacefully by in the cool of the evening. I particularly like doing maths in unusual places. Where people aren't expecting it. Where they are not on their guard . . .

'Can't wait to tell my parents I did some maths during the holidays!' This from a secondary school pupil as he walks past my stall on his way back from the beach.

It's true – I do catch them slightly unawares. But sometimes you've got to do what you've got to do. This is one of my favourite moments: observing the expression on the faces of people who thought that they had fallen out with maths for good at the instant when I tell them that they have just been doing maths for fifteen minutes. And my stall is always crowded! I present origami, magic tricks, games, riddles . . . there's something for everyone.

No matter how much this amuses me, on balance I find it upsetting. How has it come about that we need to hide from people the fact that they are doing maths before they can take some pleasure in it? Why is the word so frightening? One thing is certain: had I put up a sign above my table proclaiming 'Mathematics' as visibly as 'Jewels and necklaces', 'Phones' or 'Tattooing' on the stalls around me, I would not have had a quarter of the same success. People would not have stopped. Perhaps they would even have turned away and averted their gaze.

All the same, the curiosity is there. I observe this every day. Mathematics is scary, yet even more, it is fascinating. Some may not like it, but would like to like it, or at least to be able to peep at will into its murky mysteries. Many think it is inaccessible. But this is not true. It is perfectly possible to love music without being a musician, or to like to share a nice meal without being a great cook. Then why should you have to be a mathematician, or someone exceptionally clever, in order to be open to hearing about mathematics and to enjoy having your imagination tickled by algebra or geometry? It is not necessary to delve into the technical details in order to understand the great ideas and to be able to marvel at them.

Since time out of mind, innumerable artists, creative spirits, inventors, artisans, simple dreamers, or the purely curious, have done maths without being aware of the fact: mathematicians despite themselves. They were the first to ask questions, the first to undertake research, the first to brainstorm. If we want to understand the whys and wherefores of mathematics, we have to set out on their trail, since it all began with them.

The time has come to embark on a journey. Please allow me the space of these few pages to carry you with me into the twists and turns of one of the most fascinating and most astonishing disciplines ever practised by humankind. Let us set off to meet those who have created its story through unexpected discoveries and fabulous ideas.

Let us set out on a big adventure and see how it all adds up.

1

MATHEMATICIANS WITHOUT KNOWING IT

Back in Paris, it is at the Louvre Museum, in the heart of the French capital, that I decide to begin our investigation. Am I going to do maths in the Louvre? That may seem incongruous. Nowadays the former royal residence, converted into a museum, seems to be the domain of painters, sculptors, archaeologists and historians rather than of mathematicians. However, it is here that we can now hark back to the first traces of those same mathematicians.

From the moment I arrive, the appearance of the great glass pyramid that has pride of place in the centre of the Napoleon Courtyard is already an invitation to geometry. But today I have an appointment with a much more ancient era. As I enter the museum the time machine wakes up. I pass by the Kings of France and travel back through the Renaissance and the Middle Ages before arriving in Antiquity. The rooms spool by, I come across a few Roman statues, Greek vases, Egyptian sarcophagi. Still a little further to go. Now I'm entering prehistory, and as I speed through the centuries I gradually need to forget everything: numbers, geometry, writing. In the beginning, no one knew anything. No one even knew that there was something to know.

The first stop is in Mesopotamia. We have travelled back ten thousand years.

Come to think of it, I could have gone back even further. I could have flown back another one-and-a-half million years, to the heart of the Palaeolithic, the Old Stone Age. Fire has not yet been tamed and *Homo sapiens* is only a distant prospect. *Homo erectus* reigns in Asia, and *Homo ergaster* in Africa, perhaps with other cousins as yet undiscovered. This is the age of flake stone tools. The biface has arrived.

Flint knappers are at work in a corner of the encampment. One of them picks up a virgin piece of flint, which is still just as it was when he chose it a few hours earlier. He sits down on the bare earth, places the stone on the ground, and holds it steady with one hand, while with the other hand he strikes its edge with a heavy stone. A first flake breaks off. He examines the result, turns his flint over and strikes it a second time on the other side. The first two flakes that have broken off in this way, opposite one another, leave a sharp edge on the flint. It only remains to repeat the operation along the whole of the contour. In some places the flint is too thick or too wide, and larger pieces have to be removed in order to give the final object the shape desired.

For the shape of the biface is not left to chance or to the whim of the moment. It is designed, worked, and transmitted from generation to generation. We find various different models, according to the period and to where they were produced. Some are shaped like a drop of water with a prominent point, other more rounded ones are egg-shaped, while others are more like an isosceles triangle with very slightly curved sides.

Biface from the Lower Palaeolithic

But they all have one thing in common: an axis of symmetry. Could there be a practical aspect to this geometry, or was it simply an aesthetic intention that drove our ancestors to adopt these shapes? The flint knapper's blows had to be premeditated. He had to think about the shape before creating it, to construct an abstract image in his own mind of the object to be crafted. In other words, he had to do some mathematics.

When he finished the piece, the knapper inspected his new tool, held it up to the light at arm's length to better scrutinize the contour, and altered a few cutting edges with two or three more sharp taps before he was finally satisfied. How did he feel at that moment? Did he already experience the tremendous exaltation of scientific creativity: of having been able to grasp and shape the outside world, through an abstract idea? It doesn't matter; the finest hours of abstraction had not yet arrived. This is a time of pragmatism. The biface would be used to cut wood, to cut up meat, to pierce animal hides and to dig holes in the ground.

But no, we shall not travel so far back. Let us allow these ancient times and these few random interpretations to rest in peace, and let us return to what will become the true point of departure for our adventure: the Mesopotamian region in the eighth millennium BC.

All along the Fertile Crescent, in a zone that roughly corresponds to the area now known as Iraq, the Neolithic revolution was under way. People had been settling there for some time. In the northern plateaux, sedentarization was a success story. The region was the laboratory for all the very latest innovations. The mudbrick houses formed the first villages, and the bravest builders were even at that time adding a storey. Agriculture was a state-of-the-art technology. The temperate climate meant that the land could be cultivated without artificial irrigation. Animals and plants were gradually being domesticated. Pottery was on the verge of emerging.

Speaking of which, let's talk about pottery. For while many testimonies to these periods have disappeared, irreparably lost in the mists of time, nevertheless archaeologists continue to gather thousands of artefacts: pots, vases, jars, plates, bowls, etc. The display cabinets around me are full of them. The earliest date from 9,000 years ago, and from room to room, like the pebbles of Hop-o'-My-Thumb, they guide us through the centuries. They come in all shapes and sizes and are variously decorated, sculpted, painted or engraved. Some have feet, others have handles. Some are intact, cracked, broken, or have been reconstituted. In some cases, only a few sparse fragments remain.

Ceramics was the first art to use fire – well before bronze, iron or glass. Artisan potters were able to use clay, that paste of malleable rock and soil material which can be harvested in abundance in these humid regions, to fashion objects as they wished. When they were satisfied with the shape, they had only to let the objects dry for a few days, before cooking them in a great fire so that they solidified. This technique had already been known for a long time. Twenty thousand years earlier people were already producing small statuettes. But the idea of producing utensils out of clay had only arisen recently with sedentarization. The new way of life required a means of storage, so pots were produced in bulk.

These earthenware receptacles rapidly came to be indispensable to everyday life and also necessary for the collective organization of the village. If there was going to be washing up to do, it might as well be beautiful – soon, pots were being decorated. Here again, there were several approaches. Some potters imprinted their patterns in the clay while it was still fresh by using a shell or a twig, before baking the pot. Some did the baking first before engraving their decorations using stone tools. Others preferred to paint on the surface with the help of natural pigments.

As I pass through the rooms of the Near Eastern Antiquities section, I am struck by the richness of the geometric patterns conceived by the Mesopotamians. Just as it was for our flint knapper's biface, some symmetries are so ingenious that they must have been very thoughtfully designed in advance. The friezes running round the rims of these vases pique my interest.

The friezes are decorated bands with a single pattern repeated round the whole circumference of the pot. The most common ones include triangular sawtooth designs. There are also friezes showing two intertwined strings. Then there are friezes with angled rectangles, friezes with square indentations, friezes with pointed lozenges, with hatched triangles, with interlocking circles, and so on.

When one moves from one region to another or from one period to the next, trends can be seen. Certain patterns are very popular. They are reworked, transformed, improved in multiple variants. Then, several years later, they are essentially abandoned, replaced by other designs that are fashionable at the time.

My mathematician's eyes light up as I spot symmetries, rotations and translations. Then in my mind I begin to sort and arrange all this. Several theorems from my undergraduate days come back to me. What I need is the classification of geometric transformations. I take out a notebook and a pencil and begin to scribble.

First, there are rotations. Appropriately, in front of me there is a frieze consisting of S-shaped patterns interlocking one after the other. I tilt my head to make sure I'm right. Yes, there's no doubt, this one is invariant under a half-turn: if I were to take the earthenware jar and turn it upside down, the frieze would still look exactly the same.

Then there are symmetries. There are several types of these. I am gradually filling in my list and a treasure hunt begins. For each geometric transformation I look for the corresponding frieze. Moving from room to room, I retrace my steps. Some pieces are damaged, so I squint to try to reconstitute the patterns that ran across this clay thousands of years ago. When I find a new pattern, I tick it off. I look at the dates to try to reconstitute the chronology in which the patterns appeared.

How many patterns should I find in total? After a little thought, I finally manage to put my finger on this famous theorem. There are, in sum total, seven categories of friezes. Seven different groups of geometric transformations that can leave them invariant. Not one more, not one less.

That is not something the Mesopotamians knew. And with good reason, as the first formalization of the theory in question only began in the Renaissance. However, without knowing it, and without pretending to be doing anything other than decorating their pottery with harmonious and original designs, these prehistoric potters were actually undertaking the very first reasoning in a fantastic discipline, that would focus the efforts of a whole community of mathematicians thousands of years later.

Looking over my notes, I see almost all the patterns. Almost? I'm still missing one of the seven friezes. This doesn't surprise me a lot, as this frieze is clearly the most complicated one on the list. I am looking for a frieze that, if you flip it over horizontally, will have the same appearance but shifted by half a pattern length. Today, we call this a glide symmetry. This would have been a real challenge for Mesopotamians.

However, I still have not visited all the rooms, so I'm not losing hope – the hunt is still on. I observe the smallest detail, the least indication. Examples of the other six categories, those I have already observed, are piling up. In my notebook, the dates, sketches and other scribbles are getting muddled. But still no sign of the mysterious seventh frieze.

All at once, a rush of adrenalin. Behind this glass panel I have just spotted a piece that looks in a bit of a sorry state. A simple fragment. From top to bottom you can clearly make out four partial friezes, one above the other, and one of these has caught my eye. The third one down – it consists of what look like fragments of inclined interlocking rectangles. My eyes are blinking. I'm peering attentively, sketching the pattern quickly in my notebook, as though I'm afraid it will vanish before my eyes. It has the right geometry. This really is the glide symmetry. The seventh frieze is unmasked.

The caption next to it says: *Fragment of beaker with horizontal décor of bands and pointed lozenges. Middle of the fifth millennium* BC.

I insert it mentally in my chronology. Mid-fifth millennium BC. We are still in prehistory. More than a thousand years before the invention of writing – without knowing it – the Mesopotamian potters had listed all the cases of a theorem that would only be stated and proved 6,000 years later.

A few rooms further, I come across an earthenware jar with three handles, which also falls into the seventh category: even though the pattern has become a spiral, the geometric structure remains the same. A little further on again, there is another one. I would like to continue, but suddenly the décor changes and I have arrived at the end of the Near Eastern collection. If I continue, I will reach Greece. I cast a last glance at my notes; the friezes with glide symmetry can be counted on the fingers of one hand. I had a stroke of luck.

HOW TO RECOGNIZE THE SEVEN CATEGORIES OF FRIEZES

The first category is that of friezes that have no particular geometric property, and in which a simple pattern is repeated without symmetries or centres of rotation. This is, in particular, the case for friezes that are not based on geometric figures but on figurative drawings, for instance of animals.

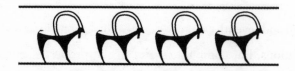

The second category consists of friezes in which the horizontal line dividing the figure in two is an axis of symmetry.

The third category comprises friezes that have a vertical axis of symmetry. Since a frieze consists of a pattern that is repeated horizontally, the vertical axes of symmetry are also repeated.

The fourth category is that of friezes that are invariant under a rotation by a half-turn. Whether you look at these friezes from above or below you will always see the same thing.

The fifth category is that of glide symmetries. This notable category was the last one I discovered while passing through Mesopotamia. If you reflect such a frieze about a horizontal axis of symmetry (as per the second category) the frieze obtained is similar, but translated by half a pattern length.

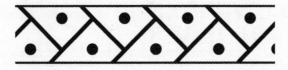

The sixth and seventh categories do not correspond to new geometric transformations, but combine several of the properties encountered in the previous categories. Thus, the friezes of the sixth category are those that have a horizontal symmetry, a vertical symmetry and a centre of rotation by a half-turn, all at the same time.

Finally, the seventh category consists of friezes that have a vertical symmetry, a centre of rotation and a glide symmetry.

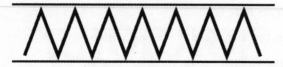

Note that these categories only relate to the geometric structure of the friezes and do not exclude possible variations in the pattern designs. Thus, while the following friezes are all different, they all belong to the seventh category.

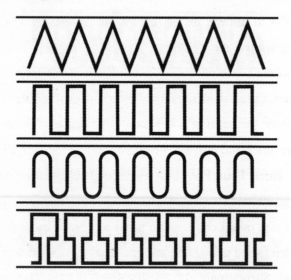

Therefore, all the friezes that you can possibly imagine belong to one of these seven categories. Any other combination is geometrically impossible. Curiously, the last two categories are the most common.

It is spontaneously easier to draw figures with a lot of symmetries than figures that have only a few.

Bursting with pride following my successes with the Mesopotamians, I am back the very next day, ready to take ancient Greece by storm. I have barely arrived and already I am overwhelmed with choice. The hunt for friezes here is child's play. After just a few steps, a few display cases and a few black amphoras decorated with red figures, I have already found my list of seven friezes again.

Faced with such an abundance of riches, I quickly decide not to record my statistics as I had done in Mesopotamia. I am amazed by the creativity of the artists. Ever more complex and ingenious new patterns come into view. I have to pause several times and concentrate hard in order to unravel in my mind all the intertwined visions that are swirling around me.

In one room, a *loutrophoros* with red figures takes my breath away. A *loutrophoros* is a tall slender vase with two handles; its function is to carry water for bathing – this one is almost one metre tall. The friezes come thick and fast, and I begin by ticking them off by category. One. Two. Three. Four. Five. In a few seconds I have identified five of the seven geometric structures. The vase is standing against the wall, but when I lean over I can see that it has a frieze of the sixth category on its hidden face. I'm missing just one frieze now. That would be too much to ask. Surprisingly, the missing one is not the same one as the day before. Times have changed, and fashions too; I'm no longer missing the glide symmetry, but the combination comprising vertical symmetry, rotation and glide symmetry.

I am looking frantically for it, my eyes scanning each and every feature of the object. But I can't find it. I'm slightly disappointed and preparing to give up when my gaze alights on a detail. In the middle of the vase there is a depiction of a scene involving two people. At first sight, there does not seem to be a frieze at that point. Then, an object at the bottom right-hand corner of the scene stops me in my tracks: a vase on which the central person is resting. A vase drawn on the vase! This technique of *mise en abyme*, in which an image contains a smaller copy of itself, makes me smile. I look closer. The image is a little damaged, but there is no doubt about it: this image of a vase itself contains a frieze and, what a miracle, it's the one that I'm missing!

Despite my repeated efforts I will fail to find another piece with the same property. This *loutrophoros* seems to be unique in the Louvre's collections: it is the only one that bears all seven categories of friezes.

A little further on, another surprise awaits me: friezes in 3D! And I had thought perspective was an invention of the Renaissance. Bright and dark areas skilfully positioned by the artist create a *chiaroscuro* effect, adding volume to the geometric shapes that pursue each other around the circumference of this gigantic receptacle.

The further I proceed, the more questions arise. Some pieces are not covered by friezes but by tilings. In other words, the geometric patterns no longer simply fill a narrow band running around the object, but now also invade its whole surface, thereby reducing the possibilities for geometric combinations.

After the Greeks come the Egyptians, the Etruscans and the Romans. I discover illusions of filigree lace carved in stone. The stone threads are interwoven and pass alternately above and below one another in a perfectly regular meshing. Then, as if the works of art are not sufficient in themselves, I am soon surprised to find myself observing the Louvre itself, its ceilings, its tiled floors and its door surrounds. As I return home, it's as though I am unable to stop myself. In the street, I look at the balconies of buildings, the patterns on the clothes worn by passers-by, the walls in the corridors of the Métro, etc.

You have only to change how you look at the world to see mathematics emerge. The search is fascinating. It has no end.

And the adventure is just beginning.

2

AND THEN THERE WERE NUMBERS

In those days, in Mesopotamia, the region was thriving. At the end of the fourth millennium BC, the small villages we visited earlier had transformed into flourishing towns. Some of them now already had several tens of thousands of inhabitants. Technology was advancing at a rate never before seen. Artisans, whether they were architects, goldsmiths, potters, weavers, carpenters or sculptors, needed to exhibit constant ingenuity to cope with the technical challenges they confronted. Metallurgy was not yet at its peak, but it was a work in progress.

Gradually, a network of routes came to criss-cross the whole region. Cultural and commercial exchanges multiplied. Increasingly complex hierarchies established themselves and *Homo sapiens* discovered the joys of administration. All this took extensive organization. To establish some degree of order, it was high time for our species to invent writing and enter history. Mathematics was to play a cutting-edge role in this nascent revolution.

Following the course of the Euphrates, we leave the northern plateaux that saw the birth of the first sedentary villages, and head towards the Sumer region that covers the plains of Lower Mesopotamia. It is here, in the southern steppes, that the chief population centres were then concentrated. Along the river we reach the towns of Kish, Nippur and Shuruppak. These towns were still young, but the centuries that lay ahead of them had grandeur and prosperity in store.

Then, suddenly, Uruk emerges.

The town of Uruk was a human anthill, which lit up the whole of the Near East with its prestige and power. It was built mainly of mud bricks, and its orangey hues extended over more than one hundred hectares, so that the new visitor could stray for hours in its crowded streets. At the heart of the town, several monumental temples had been constructed, dedicated to An, the father of all the gods, and above all to Inanna, the Lady of the Heavens. It was for her that the Eanna Temple was built, its largest building 80 metres long by 30 metres wide, and hugely impressive to the many travellers who were drawn here.

Summer was coming, and as in every year at this time, the whole town was beginning to bustle with a particular activity. Soon the flocks of sheep would leave for the northern pasturelands, returning only at the end of the hot season. For several months, it would be the job of the shepherds to manage their animals, to keep them fed and watered, and to return them safe and sound to their owners. The Eanna Temple itself owned several flocks, the largest numbering several tens of thousands of animals. The convoys were so impressive that some were escorted by soldiers to protect them from the dangers of the journey. However, for the owners it was out of the question to let their sheep go without a few precautions. As for the shepherds, the contract was clear: as many animals must return as had set out. There was no margin for losing part of the herd or for clandestine trading.

This then led to a problem: how do you compare the size of the flock that leaves with the size of the one that returns?

In response to this, a system of clay tokens had already been developed several centuries earlier. There were several types of token, each representing one or more objects or animals according to its shape and the patterns traced on it. For a sheep, there was a simple disc marked with a cross. At the time of leaving, a number of tokens corresponding to the size of the flock were placed in a receptacle. On return, the owners had only to compare the flock with the contents of the receptacle to check that no animal was missing. Much later on, these tokens were given the Latin name of *calculi*, 'small stones', from which the term *calculus* was derived.

This method was practical, but it had a disadvantage. Who looked after the tokens? For suspicion cuts both ways, and the shepherds were in turn worried that unscrupulous owners might add extra tokens to the urn during their absence. Then they could claim compensation for non-existent sheep.

After much racking of brains, a solution was found. The tokens were to be kept securely in a sealed hollow clay sphere or envelope. At the time it was sealed, both parties were to leave their signatures on the surface of the sphere to certify its authenticity. It was then impossible to modify the number of tokens without breaking the sphere. The shepherds could leave without worrying.

But then, once again, it was the owners who found drawbacks in this method. For their business requirements, they needed to know the number of animals in their flocks at any time. How could this be achieved? Could the number of sheep be committed to memory? This was not straightforward, since the Sumerian language did not yet have words to denote such large numbers. Could you have an unsealed duplicate of all the tokens contained in all the envelopes? This was not very practical.

At last a solution was found. A reed stem was cut and used to trace out on the surface of each sphere a representation of the tokens it contained. That made it possible to reference the contents of the envelope at will without having to break open the sphere.

This method now seemed to suit everyone. It was widely used, not only to count sheep, but also to set a seal on all kinds of agreements. Cereals such as barley or wheat, wool and textiles, metal, jewels, precious stones, oil and pottery also had their tokens. Even taxes were controlled by tokens. In short, at the end of the fifth millennium, in Uruk, any contract in due and proper form had to be sealed by a hollow-sphere envelope with its clay tokens.

All this worked wonderfully well, but then one day a new idea emerged, so brilliant and so simple that one wonders why no one had thought of it before. Since the number of animals was inscribed on the surface of the sphere, what was the point of continuing to put the tokens inside it? And what was the point of continuing to make spheres? You could simply draw a representation of the tokens on an arbitrary piece of clay – for example, on a flat tablet.

This came to be called writing.

I'm back in the Louvre. The collections of the department of Near Eastern Antiquities bear witness to this story. The first thing that strikes me when I see these sphere–envelopes is their size. These clay spheres that the Sumerians created simply by turning them around their thumbs are scarcely any larger than ping-pong balls. As for the tokens, they are no bigger than a centimetre.

Stepping further into the museum brings us to the first tablets. Their numbers grow and they quickly come to fill whole display slots. Over time, the writing became more precise and took on its cuneiform appearance, comprising small notched wedges in the shape of a nail, with a stem and a head. Following the disappearance of the first Mesopotamian civilizations at the beginning of the modern era, most of these pieces had lain quietly for centuries beneath the ruins of those deserted towns before they were unearthed by European archaeologists from the seventeenth century onwards. They were only gradually deciphered during the nineteenth century.

These tablets are not very large either. Some of them are the size of simple visiting cards, but covered with hundreds of tiny signs which are crammed together one above the other. Mesopotamian scribes did not waste the slightest portion of the clay when they wrote. The museum's explanation cards placed alongside these pieces enable me to interpret these mysterious symbols. They concern livestock, jewels or cereals.

Next to me, some tourists are taking photos with their tablets – this is an amusing nod to the carousel of history where writing appears on so many different media, from clay to paper, by way of marble, wax, papyrus and parchment, and which, in a final jest, has endowed electronic tablets with the shape of their earthenware ancestors. The face-to-face encounter of these two objects has something particularly touching about it. For all we know, in five thousand years' time these two tablets may find themselves next to each other again, but on the same side of the glass.

Time has passed. We are now at the start of the third millennium BC, and another step has been taken: numbers have been freed from the objects that they count. Previously, in the case of sphere-envelopes and the very first tablets, the counting symbols depended on the objects in question. A sheep is not a cow, so the symbol for counting a sheep was not the same as the one for counting a cow. Every object that could be counted had its own symbols, just as it had had its own tokens.

But all that was over and done. Numbers had acquired their own symbols. In other words, eight sheep were no longer counted using eight symbols each denoting a sheep, but instead the figure eight was written down followed by the symbol for a sheep. And to count eight cows, you simply had to replace the symbol for a sheep by that for a cow. The number itself stayed the same.

This stage in the history of thought was absolutely fundamental. If I had to assign an actual date to the birth of mathematics, I would be sure to choose this moment: the moment when numbers came into existence in their own right, when numbers became detached from reality to observe it from a greater height. Everything before that had simply amounted to a gestation period. Bifaces, friezes, tokens – all these amount to preludes in the scheme that led towards the birth of numbers.

From that point on, numbers became an abstract concept, and that is what gave mathematics its identity: it is the science of abstraction par excellence. The objects that mathematics studies do not have a physical existence. They are not material, they are not made of atoms, they're purely ideas. And yet these ideas are remarkably effective when it comes to making sense of the world.

It is undoubtedly no accident that the need to be able to write down numbers acted at the time as a determining factor in the emergence of writing. While other ideas could be transmitted orally without any problem, it seemed on the other hand difficult to establish a system of numbers without moving to a written notation.

Can we, even today, dissociate the idea we have of numbers from our written representation of them? If I ask you to think of a sheep, how do you see it? You probably represent it as a bleating animal with four legs and a woolly back. You would never think of visualizing the five letters of the word 'sheep'. Yet if I talk now about the number one hundred and twenty-eight, what do you see? Do the 1, the 2 and the 8 take shape in your brain and come together as though they are written in the intangible ink of your thoughts? Our mental representation of large numbers appears to be definitively linked to their written form.

This example came out of the blue. Although for everything else, writing was just a means of transcribing something that had previously existed in the spoken language, now, in the case of numbers, it was the writing that would dictate the language. Think about it: when you pronounce 'One hundred and twenty-eight', you are just reading 128: 100 + 20 + 8. Above a certain threshold, it becomes impossible to speak about numbers without the written medium. Before they were written down, there were no words for large numbers.

To this day, some indigenous peoples still only have a very limited number of words for designating numbers. For example, the members of the Pirahã tribe of hunter-gatherers who live on the banks of the Maici River in Amazonia only count to two. Above that, they use a single word signifying 'several' or 'many'. Again in

Amazonia, the Munduruku only have words for up to five – that is to say a hand.

In our modern societies, numbers have invaded our everyday lives. They have become so omnipresent and indispensable that we often forget just how brilliant the idea is, and that it took our ancestors centuries to provide us with evidence of this.

Throughout the ages, very many procedures for recording numbers have been invented. The simplest of these involves writing down as many signs as the desired number – for example, a sequence of bars side by side. We often still use this method, for example to count the points in a game.

The oldest known indication of the probable use of this procedure dates back to well before the invention of writing by the Sumerians. The so-called Ishango bone (actually two baboon bones) was found in the 1950s on the shores of Lake Edward in what is now the Democratic Republic of Congo and dates back some twenty thousand years. The bones are 10 and 14 centimetres long, and feature a large number of more or less regularly spaced incisions. What was the role of these incisions? It is likely that this was a first counting system. Some people think the bones are a calendar, while others extrapolate an already highly advanced knowledge of arithmetic. It is hard to know for sure. The two bones are now on display in the museum of the Royal Belgian Institute of National Sciences in Brussels.

This method of counting using a mark for each unit added rapidly reached its limits as soon as there was a need to manipulate relatively large numbers. Packets were introduced to achieve a greater speed.

The Mesopotamian tokens could already represent several units. For example, there was a particular token for representing ten sheep. This principle was retained at the time that writing came in. There are symbols to denote packets of 10, of 60, of 600, of 3,600 and of 36,000.

Here we also see the search for a logic in the construction of the symbols. For example, the 60 and the 3,600 are multiplied by 10 when a circle is included inside them. With the arrival of cuneiform writing, these first symbols were gradually transformed into the following:

Because it was so close to Mesopotamia, Egypt did not lag far behind in adopting writing, and developed its own numeration symbols from the start of the third millennium.

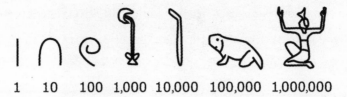

| 1 | 10 | 100 | 1,000 | 10,000 | 100,000 | 1,000,000 |

The system was then purely decimal: each symbol had a value ten times greater than the one before.

These additive systems, in which one just has to add the values of the written symbols, became very popular throughout the world, and numerous variants were unveiled throughout antiquity and also much of the Middle Ages. They were used, in particular, by the Greeks and the Romans, who simply employed the letters of their respective alphabets as numerical symbols.

Alongside the additive systems, a new form of notation for numbers gradually emerged: numeration by position. In these systems, the value of a symbol was defined to depend on the position it occupies within the number. Once again, the Mesopotamians were the first to come up with this.

In the second millennium BC, it was the town of Babylon that now shone brightly over the Near East. Cuneiform writing was still in vogue, but now, just two symbols were used: the simple nail that had the value '1' and the chevron with the value '10'.

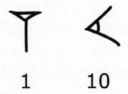

| 1 | 10 |

These two signs were used to provide an additive notation for all numbers up to 59. For example, the number 32 was written as three chevrons followed by two nails.

32

Then, from 60 onwards, groups were introduced, where the same symbols were used to denote groups of 60. Thus, in the same way as in our present-day notation, where the figures read from right to left denote the units, then the tens, then the hundreds, in this Babylonian numeration the units are read first, then the sixties, then the three-thousand-six-hundreds (that is, sixty sixties), and so on, where each rank has a value sixty times greater than its predecessor.

For example, the number 145 consists of two sixties (which make 120) to which you have to add 25 units. The Babylonians would therefore have denoted it as follows:

2   sixties   25   units

145

Based on this system, Babylonian scholars developed an extraordinary knowledge. They had a good understanding of the four basic operations of addition, subtraction, multiplication and division, and also of square roots, powers and reciprocals. They produced extremely detailed arithmetic tables and developed very good solution techniques for equations that they set themselves.

However, all this was soon forgotten. The Babylonian civilization was in decline, and a large part of its mathematical advances would be consigned to oblivion. There would be no more numeration by position and no more equations. Indeed, there was a time lag of centuries before these questions became the flavour of the day again, and it was only in the nineteenth century that the decoding of cuneiform tablets reminded us that the Mesopotamians had tackled these things before everyone else.

Following the Babylonians, the Mayans also devised a positional system, in this case with base 20. Then it was the turn of the Indians to invent a system with base 10. This last system was used by Arab scholars before it reached Europe at the end of the Middle Ages. Its symbols became known as Arabic figures, and soon spread all over the world.

0 1 2 3 4 5 6 7 8 9

With numbers, mankind gradually came to understand that it had just invented a tool for describing, analysing and understanding the world around it that surpassed any purpose it might have hoped for.

Sometimes we have been so pleased with numbers that we have even overdone it. The birth of numbers represented at the same time the birth of the practice of various forms of numerology. This involves attributing magical properties to numbers, interpreting them beyond the bounds of reason, and attempting to read into them messages from the gods and about the fate of the world.

In the sixth century BC, the Greek scholar Pythagoras made numbers the fundamental concept of his philosophy when he declared: 'Everything is a number.' According to him, it is numbers that produce geometric figures, which in turn give rise to the four elements of matter – fire, water, earth and air – of which all living things are made. Pythagoras thus created a whole system around numbers. The odd numbers were associated with the masculine, while the even numbers were feminine. The number 10, represented as a triangle called the *tetractys*, became a symbol of harmony and of the perfection of the cosmos. Pythagoreans were also behind the origin of arithmancy, which claimed to read people's characters by associating numerical values with the letters comprising their names.

In parallel, there began to be discussions about what constitutes a number. Some authors believed that the unit – 1 – is not a number, because a number denotes a plurality and so can only be considered from 2 onwards. It was even asserted that in order for it to be able to generate all the other numbers, 1 must be simultaneously even and odd.

Later, increasingly animated discussions developed concerning the zero, the negative numbers and the imaginary numbers. In each case, the admission of these new ideas to the circle of numbers led to debate and forced mathematicians to broaden their ideas.

In short, numbers have never ceased to raise questions, and human beings still need time to learn to master these strange creatures that are their brainchildren.

3

LET NO ONE IGNORANT
OF GEOMETRY ENTER

Once numbers had been invented, it did not take long for the discipline of mathematics to spread its wings. Various core branches such as arithmetic, logic and algebra gradually sprouted within it, developed to maturity and asserted themselves as disciplines in their own right.

One of these, geometry, rapidly won the popularity stakes and captivated the greatest scholars of antiquity. It was this that singled out the first celebrities of mathematics, such as Thales, Pythagoras and Archimedes, whose names still haunt the pages of our textbooks.

However, before it became a subject for great minds, geometry gained its place on the ground. Its etymology bears witness to this: it is first and foremost the science of the measurement of the Earth, and the first surveyors were hands-on mathematicians. Problems concerning the division of territory were then classics of the craft. How to divide a field into equal parts? How to determine the price of a plot of land from its area? Which of two plots is closer to the river? What route should the future canal follow to make it the shortest possible?

All these questions were paramount in ancient societies where the whole economy still revolved in a vital way around agriculture and hence around the distribution of land. In response to this, geometrical know-how was built up, enriched and transmitted from generation to generation. Anyone equipped with this know-how was certain to hold a central and indisputable place in society.

For these measurement professionals, the rope was often the primary instrument of geometry. In Egypt, 'ropestretcher' was a profession in its own right. When the Nile floods led to regular inundations, it was the ropestretchers who were sent for to redefine the boundaries of plots that bordered the river. Using information they recorded about the ground, they planted their stakes, stretched their long ropes across the fields, and then carried out calculations that enabled them to rediscover the boundaries erased by the floodwaters.

They were also the first port of call in constructing buildings, when they took the measurements on the ground and marked the precise location of the building based on architects' plans. And in the case of a temple or an important monument, it was often the pharaoh in person who symbolically came to stretch the first rope.

It can be said that the rope was the all-purpose tool of geometry. Surveyors used it as a ruler, as compasses, and as a set-square.

To use it as a ruler is straightforward: if you stretch the rope between two fixed points you obtain a straight line. And if you require a grad-uated ruler, you just tie knots at regular intervals along your rope. For compasses, there is no magic involved either. You simply fix one of the two ends to a stake, stretch the rope and move the other end around the stake. This gives a circle. And if your rope is graduated, you can control the length of the radius exactly.

For the set-square, however, things are slightly more complicated. Let's look at this particular problem for a few moments: what would you do to draw a right angle? With a bit of research, one can come up with several different methods. If, for example, you draw two circles that intersect each other, then the straight line that joins their centres is perpendicular to the straight line that passes through their two points of intersection. There is your right angle.

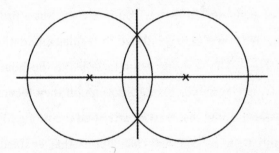

From a theoretical point of view, this construction works perfectly, but things are more complicated in practice. Imagine the surveyors, out in the fields, having to lay out two large circles precisely every time they needed to draw a right angle or, more simply, to verify that an angle which had already been constructed was actually a right angle. This was neither fast nor efficient.

The surveyors adopted a different method, which was subtler and more practical: they used their rope directly to form a triangle with a right angle (known as a right-angled triangle). The most famous one is the 3–4–5. If you take a rope divided into twelve intervals by thirteen knots, then you can form a triangle whose sides measure three, four and five intervals, respectively. And, as if by magic, the angle formed by the sides of length 3 and 4 is a perfect right angle.

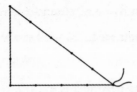

Four thousand years ago, the Babylonians already had tables of numbers that could be used to construct right-angled triangles. The Plimpton 322 Tablet, which is currently in the collections of Columbia University in New York City, and dates from 1800 BC, contains a table of fifteen triples of such numbers (so-called Pythagorean triples). Apart from the 3-4-5, it has fourteen other triangles, some of which are considerably more complicated, such as the 65-72-97 or even the 1,679-2,400-2,929. Up to a few minor mistakes, such as errors in calculation or transcription, the triangles of the Plimpton Tablet are perfectly exact, and they all have a right angle.

It is difficult to know the precise period from which the Babylonian surveyors began to use their knowledge of right-angled triangles on the ground, but the use of these triangles persisted well beyond the disappearance of the Babylonian civilization. In the Middle Ages, for example, the rope with thirteen knots remained an essential tool for cathedral builders.

On our journey through the history of mathematics, it is by no means uncommon to find certain similar ideas appearing independently at removes of thousands of kilometres and in profoundly different cultural contexts. One such startling coincidence is that during the first millennium BC the Chinese civilization developed a whole mathematical know-how that corresponds remarkably to that of the Babylonian, Egyptian and Greek civilizations of the same period.

This knowledge was amassed over the centuries before being

compiled under the Han dynasty, around 2,200 years ago, into one of the world's first great mathematical works: *The Nine Chapters on the Mathematical Art.*

The first of these *Nine Chapters* is entirely devoted to the study of measurements of fields of various shapes. Rectangles, triangles, trapezia, circles, portions of circles and also rings represent geometric figures for which procedures for calculating their areas are described in minute detail. Later in the work, you discover that the ninth and final chapter deals with right-angled triangles. And guess which figure is discussed from the very first sentence of this chapter: the 3-4-5!

Good ideas are like that. They transcend cultural differences and are able to blossom spontaneously wherever human minds are ready to devise and absorb them.

SOME PROBLEMS OF THE PERIOD

Questions about fields, about architecture, or more generally about land planning led the scholars of antiquity to set themselves a great diversity of geometric problems of which the following are examples.

The following statement, taken from the Babylonian Tablet BM 85200, shows that the Babylonians were not content with plane geometry, but also thought in the space dimension.

An excavation. So much as the length, that is the depth. 1 the earth I have torn out. My ground and the earth I have accumulated, 1' 10. Length and width: '50. Length and width, what?*

As you will have gathered, the style of the mathematicians of Babylon

---

* Taken from: Jens Høyrup, *Lengths, Widths, Surfaces*, Springer-Verlag, 2010.

was telegraphic in nature. If we expand it further, this same statement might look as follows:

*The depth of an excavation is twelve times greater than its length.** *If i dig further so that my excavation has one more unit of depth, then its volume will be equal to 7/6. If i add the length to the width i obtain 5/6.*[†] *What are the dimensions of the excavation?*

This problem was accompanied by the detailed method for solving it, ending with the solution: the length is ½, the width ⅓ and the depth 6.

Let's now take a short trip down the Nile. As a matter of course, in the case of the Egyptians, we find problems about pyramids. The following statement is an extract from a famous papyrus, the Rhind papyrus, copied by the scribe Ahmes, dating from the first half of the sixteenth century BC.

A pyramid has a base side of 140 cubits and an inclination[‡] *of 5 palms and 1 digit, what is its altitude?'*

The cubit, the palm and the digit were units of measurement of 52.5 centimetres, 7.5 centimetres and 1.88 centimetres, respectively. Ahmes also gave the solution: 93 + ⅓ cubits. In this same papyrus, the scribe also took on the geometry of the circle.

Example of the calculation of a round field with a diameter of 9 khet. *What is the value of its area?*

---

* The statement on the tablet appears to say that the length and the depth are equal, but in the Babylonian system depths are measured by a unit twelve times greater than that used for lengths.

† Note also that with the system to base 60, the notation 1′10 denotes the number equal to 'one plus ten sixtieths', which we denote in our own present system by the fraction 7/6. The notation ′50 denotes the fraction 5/6 (or fifty sixtieths).

‡ The inclination of the face of a pyramid, also known as *seked* in Egyptian, corresponds to the horizontal distance between two points whose height differs by a cubit.

The *khet* is also a unit of measurement representing approximately 52.5 metres. To solve this problem, Ahmes stated that the area of this circular field is equal to that of a square field with a side of length 8 *khet*. The comparison is extremely useful, for it is much easier to calculate the area of a square than that of a circle. His solution was $8 \times 8 = 64$. However, the mathematicians who succeeded Ahmes came to discover that his result was not quite exact. The areas of the circle and the square do not quite agree. Since then, many people have attempted to answer the question: how do you construct a square with an area equal to that of a given circle? Many have worn themselves out in vain in this pursuit, but for a reason. Without knowing it, Ahmes was one of the first to tackle what would become one of history's classic mathematical conundrums: the squaring of the circle.

In China too, people sought to calculate the area of circular fields. The following problem is taken from the first of the *Nine Chapters*.

Suppose one has a circular field with circumference 30 bu *and diameter 10* bu. *The question is how big is the field?**

Here, a *bu* is equal to about 1.4 metres. And, as in Egypt, the mathematicians kept tripping over the rug with this figure. The original statement was already known to be false, since a circle of diameter 10 has a circumference slightly greater than 30. However, that did not prevent Chinese scholars from putting an approximate value on the area (75 *bu*), or from complicating the task further for themselves by continuing on to questions about circular rings!

---

* Translated based on Karine Chemla and Shuchun Guo, *Les neuf chapitres*, Éditions Dunod, 2005.

Suppose one has a field in the shape of a ring with internal circumference 92 bu, external circumference 122 bu, and transverse diameter 5 bu. The question is: how big is the field?

It seems likely that there were never any ring-shaped fields in ancient China – these latter problems suggest that the scholars of the Central Kingdom were into geometry and raised these questions as purely theoretical challenges. Research into ever more improbable and weird-looking geometric figures in order to study and understand them remains a favourite pastime of our mathematicians to this day.

Among the ranks of professional geometers, one must also include the Bematists. While it was the job of surveyors and other rope-stretchers to measure fields and buildings, the Bematists had a much grander view of things. In Greece, it was the job of these men to measure long distances by counting their steps.

And sometimes their work could take them a long way from home. For example, in the fourth century BC, Alexander the Great took several Bematists with him on his campaign in Asia, which led him as far as the boundaries of what is now India. These walking geometers thus had to measure routes of several thousand kilometres in length.

Step back a little and imagine for a moment the strange spectacle of these men walking in quick time, traversing the immense landscapes of the Middle East. See them cross the plateaux of Upper Mesopotamia; walk through the arid yellow settings of the Sinai Peninsula to reach the fertile banks of the Nile Valley; then turn back, setting off to brave the mountainous massifs of the Persian Empire and the deserts of what is now Afghanistan. Can you see them, calmly walking on and on, with a dry and monotonous rhythm, and passing through the foothills of the gigantic mountains of the Hindu Kush to return via the shores of the Indian Ocean? And all the while they counted their paces tirelessly.

The image is captivating, and the outrageous challenge of their undertaking seems insane. And yet, their results were remarkably accurate: less than 5 per cent difference on average between their measurements and the true distances we know today. Alexander's Bematists thus made it possible to describe the geography of his kingdom in a way that had never before been achieved for so vast a region.

Two centuries later, in Egypt, a scholar of Greek origin named Eratosthenes conceived an even larger project – that of measuring nothing less than the circumference of the Earth. Of course, there was no question of sending poor Bematists on a tour of the planet. However, based on skilful observations of the difference in the inclination of the sun's rays between the towns of Syene (the present-day Aswan) and Alexandria, Eratosthenes calculated that the distance between the two towns must represent one fiftieth of the total circumference of the Earth.

Naturally, he then called upon the Bematists to undertake the measurement. Unlike their Greek counterparts, the Egyptian Bematists did not count their own steps, but those of a camel they took with them. The animal is renowned for the regularity of its walking. After long days of travelling along the Nile, the verdict was issued: the two towns were 5,000 stadia apart and the girth of our planet was therefore 250,000 stadia, or 39,375 kilometres. Once again, the result was astonishingly accurate, as we know today that this circumference measures exactly 40,008 kilometres. The error was less than 2 per cent.

Perhaps more than any other ancient people, the Greeks gave

geometry pride of place in the heart of their culture. It was recognized for its rigour and its ability to form the mind. For Plato, it was an obligatory rite of passage for any budding philosopher, and legend has it that the slogan 'Let no one ignorant of geometry enter' was engraved at the entrance to his academy.

Geometry was so much in vogue that it eventually spilled over into other disciplines. The arithmetical properties of numbers were thus interpreted in geometric language. See, for example this definition by Euclid extracted from the seventh book of his *Elements of Mathematics* dating from the third century BC:

And when two numbers multiplying one another make some other number then the number so created is called plane, and its sides are the numbers which multiply one another.

If I take the product 5 × 3, the numbers 5 and 3 are thus called, according to Euclid, the 'sides' of the multiplication. Why is this? It is simply because a multiplication can be represented as the area of a rectangle. If the latter has a width equal to 3 and a length of 5, its area is 5 × 3. The numbers 3 and 5 are precisely the sides of the rectangle. The result of the multiplication, 15, is for its part called the 'plane', since it corresponds geometrically to a surface.

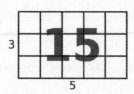

Similar constructions relate to other geometric figures. For example, a number is said to be triangular if it can be represented in the form of a triangle. The first triangular numbers are 1, 3, 6 and 10.

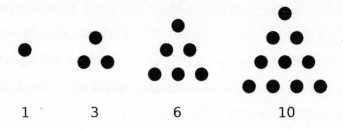

The last of the above triangles with ten points is nothing other than the famous *tetractys* that Pythagoras and his disciples employed as the symbol of the harmony of the cosmos. On the same principle, we also find the square numbers whose first representatives are 1, 4, 9 and 16.

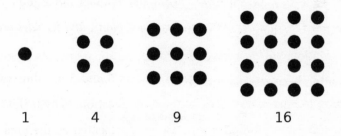

And we could of course continue for a long time like that with all kinds of figures. Thus, the geometric representation of numbers makes it possible to render visual and evident properties that are otherwise seemingly incomprehensible.

By way of example, have you ever tried adding the odd numbers successively, one after another: $1 + 3 + 5 + 7 + 9 + 11 + \ldots$? No? Well, something completely surprising happens. Look:

$$1$$
$$1 + 3 = 4$$
$$1 + 3 + 5 = 9$$
$$1 + 3 + 5 + 7 = 16$$

Can you see something special about the numbers that emerge? In the order in which they occur: 1, 4, 9, 16 . . . These are the square numbers!

And you can continue as long as you wish; this rule will never be broken. Add the first ten odd numbers from 1 to 19, if you are brave enough, and you will reach 100, which is the tenth square:

$$1 + 3 + 5 + 7 + 9 + 11 + 13 + 15 + 17 + 19$$
$$= 10 \times 10 = 100$$

Isn't that surprising? But why? What kind of miracle makes this property always hold? Of course, it would be possible to give a numerical proof, but there is a much simpler way. Using the geometrical representation, you just have to slice up the square numbers and the explanation becomes plain to see.

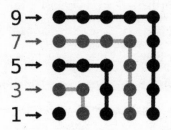

Each slice adds an odd number of balls, and increases the side of the square by one unit. This completes the simple and lucid proof.

In the kingdom of mathematics, geometry was the queen, and no assertion could be verified without coming under its scrutiny. Its reign lasted far beyond antiquity and the Greeks. It was almost two thousand years before the scholars of the Renaissance launched a vast movement to modernize mathematics which would dethrone geometry in favour of a completely new language: that of algebra.

4

THE AGE OF THEOREMS

It is the beginning of May. At midday the sun is shining over the Parc de la Villette in northern Paris. Opposite me towers the Cité des Sciences et de l'Industrie (City of Science and Industry), the largest science museum in Europe, with its Geode in the foreground. Built in the mid-1980s, this strange cinema of some 36 metres in diameter resembles a gigantic sphere with many facets.

This is very much a stopping-off place. There are tourists clutching their cameras – they have come to see this curious building in the French capital. There are families out for their Wednesday stroll. A few lovers are sitting on the grass or walking hand in hand. Here and there, a jogger zigzags through the tide of local residents who pass through with a show of indifference, scarcely casting a glance towards the strange apparition of this shimmering sphere in the middle of their day-to-day lives. All around the sphere, children amuse themselves by observing the distorted image of the surrounding world that it reflects.

As for me, if I am here today it is because I am especially interested in its geometry. I start to walk closer, scanning it closely. Its surface comprises thousands of triangular mirrors assembled together. At first sight the assembly may appear perfectly regular, but after a few minutes' scanning I begin to notice several irregularities. At certain well-defined points, the triangles grow distorted and are stretched as if by a malformation of the structure. Although they are grouped together six per hexagon, forming a perfectly regular meshing almost everywhere on the sphere, there are a dozen special points where these triangles are only grouped in fives.

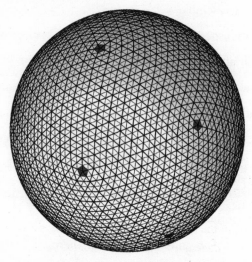

Representation of the Geode and its thousands of triangles. The points where the triangles occur in groups of five are shown in dark grey.

These irregularities are almost invisible at first glance. The fact is that most passers-by pay no attention to them. At the same time, as a mathematician, I do not see anything surprising in them. I even have to say that I expected to find them. The architect did not make a mistake; moreover, there are numerous other buildings elsewhere in the world with a similar geometry, and all have these same dozen points where the basic pieces are grouped together in fives rather than sixes. These points are the results of inescapable geometric constraints discovered over 2,000 years ago by the Greek mathematicians.

Theaetetus of Athens was a mathematician of the fourth century BC who is generally credited with the complete description of regular polyhedra. A polyhedron, in geometry, is simply a three-dimensional figure delimited by several plane faces. For example, cubes and pyramids belong to the family of polyhedra, unlike spheres and cylinders, which have rounded faces. The geode, with its triangular faces, can also be considered as a giant polyhedron, even though its large number of faces makes it resemble a sphere from a distance.

Theaetetus was especially interested in perfectly symmetric polyhedra, those for which all the faces and all the angles are the same. And his discovery was at least disconcerting: he only found five, and proved that there are no others. Five solids and that's all – not a single one more.

From left to right: the tetrahedron, the hexahedron, the octahedron, the dodecahedron and the icosahedron.

To this day, the polyhedra are usually named according to the number of their faces, written in Ancient Greek, followed by the suffix *-hedron*. For example, in geometry, the cube with its six square faces is referred to as the hexahedron. The tetrahedron, the octahedron, the dodecahedron and the icosahedron have four, eight, twelve and twenty faces, respectively. These five polyhedra subsequently became known as the Platonic solids.

Platonic? Why are they not associated in name with Theaetetus? History is sometimes unfair, and the discoverers are not always those who receive the honours of posterity. The Athenian philosopher Plato had nothing to do with the discovery of the five solids, but he made them famous through a theory that associated them with the elements of the cosmos: fire is associated with the tetrahedron, earth with the hexahedron, air with the octahedron and water with the icosahedron. As for the dodecahedron, with its pentagonal faces, Plato claimed that it was the shape of the Universe. This theory was abandoned by science a long time ago, and yet these five regular polyhedra are still conventionally associated with Plato.

As it happens, Theaetetus himself was not actually the first to discover these five solids. There exist considerably older sculpted models and written descriptions of them. For example, a collection of small sculpted stone balls reproducing the shapes of the Platonic solids came to light in Scotland and is said to date from one thousand years before Theaetetus. These pieces are currently conserved in the Ashmolean Museum in Oxford.

So is Theaetetus no more worthy than Plato? Is he also an impostor? He is not a total impostor, for while the five figures were known before his time, he was the first to demonstrate clearly that the list was complete. There is no point in looking any further, Theaetetus tells us, no one will ever find any more. This assertion has something reassuring about it. It removes a dreadful doubt. Phew – they're all present and correct!

This step is emblematic of the way in which Greek mathematicians came to approach mathematics. For them, it was no longer a matter of simply finding solutions that worked. They wanted to exhaust the problem. They wanted to make sure that nothing escaped them. And for that reason, they developed the art of mathematical exploration to its peak.

Let us now return to our Geode. Theaetetus' proof is final: it is impossible for a polyhedron with several hundred faces to be perfectly regular. But then what do you do when you are an architect and you want to create a building that resembles a regular sphere as closely as possible? It is technically difficult to design the building in a single piece. No, there's nothing you can do about that, and you have to assemble a multitude of small faces. But how do you create such a structure?

We can imagine various solutions. One of these involves taking one of the Platonic solids and modifying it. Consider the icosahedron, for example. With its twenty triangular faces, it seems the roundest of the five. To smooth it out, each of its faces can be subdivided into several smaller faces. The polyhedron obtained in this way can then be deformed, as though one were inflating it by blowing into it, in order to come closer to a sphere.

Here, for example, is what happens when you subdivide each face of the icosahedron into four smaller triangles.

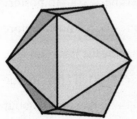  

The icosahedron   *Icosahedron with faces subdivided into four*   *Inflated icosahedron with subdivided faces*

In geometry, such a polyhedron is called a geode. Etymologically, this means a figure that is shaped like the Earth – that is to say, it resembles a sphere. There is nothing very complicated about this principle. It is precisely this construction that was used for the Geode of la Villette! The subdivision of the faces is, however, very much finer: the basic triangles of the icosahedron are here subdivided into 400 smaller triangles, making a total of eight thousand triangular facets.

In reality, the Geode has rather fewer than 8,000 facets – only 6,433 – since it is not complete. Its base, which rests on the ground, is truncated and some triangles are missing. Nevertheless, this structure allows us to explain the presence of the twelve irregularities. The latter correspond simply to the twelve vertices of the basic icosahedron. In other words, these are the points where the original large triangles met in fives to form the points of the icosahedron. These vertices, which were originally pointed, have been flattened with the increase in the number of faces to the point of becoming almost invisible. However, their presence remains firmly fixed in the layout of the triangles, and the twelve irregularities are there to remind attentive passers-by of this.

Theaetetus probably didn't imagine that his studies would one day permit the construction of buildings such as the Geode. And the great power of mathematics, as in the kind that the scholars of ancient Greece developed, lies in its formidable ability to engender new ideas. The Greeks gradually began to detach their questioning from concrete problems and thus to generate original and inspiring models out of simple intellectual curiosity. Although they might often have seemed to have no concrete use at the time they were conceived, these models sometimes turned out to be astonishingly useful long after their creators had disappeared.

Nowadays, we come across the five Platonic solids in different contexts. For example, they are well suited for use as dice in board games. Their regularity ensures that the dice are balanced, that is to say that all the faces have the same chance of occurring. Everyone is familiar with cubic dice with six faces, but the most inveterate players know that numerous other games also use the other four shapes to vary the enjoyment and the probabilities.

As I move away from the Geode, a little further on I come across some children who have produced a ball and are starting a makeshift game of football on the lawns of la Villette. They are not aware of it, but at this moment they too owe a very great deal to Theaetetus. Have they noticed that their ball also has its geometric patterns? Most footballs are formed on the same model: twenty hexagonal pieces (six-sided) and twelve pentagonal pieces (five-sided). On traditional balls, the hexagons are white while the pentagons are black. And even when the surface of the ball carries all kinds of printed inscriptions, you only have to look carefully at the seams between the different pieces to see once again the inevitable twenty hexagons and twelve pentagons.

A truncated icosahedron – that is the name geometers give to the football. And its structure is owed to the same constraints as the Geode: it has to be as regular and as round as possible. The only thing is that, in order to achieve this result, the creators of this model used a different method. Instead of subdividing the faces in order to be able to round off the corners, they simply chose to cut off the corners. Imagine an icosahedron in modelling clay, take a knife and simply slice off the vertices. The twenty triangles with their points sliced off become hexagons, while the twelve points removed lead to the appearance of the twelve pentagons.

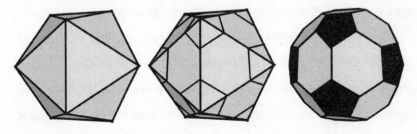

The twelve pentagons on a football thus have the same origin as the twelve irregularities on the surface of the Geode: they are the original locations of the twelve vertices of the icosahedron.

As I leave the Parc de la Villette, I come across a girl with a handkerchief. What is wrong? I wonder. She doesn't seem to be doing too well. Perhaps she has fallen victim to a nasty attack of micro-icosahedra? Certain microscopic organisms such as viruses naturally assume the form of icosahedra or dodecahedra. This is the case with the rhinoviruses, which are responsible for most colds.

If these tiny creatures adopt such forms, it is for the same reasons for which we use these forms in architecture or for our footballs, namely for the sake of symmetry and economy. Thanks to icosahedra, balls consist of only two different types of pieces. In the same way, the virus membrane consists only of a few different types of molecules (four for the rhinoviruses) that fit together by always repeating the same pattern. The genetic code needed to create such an envelope is thus very much more concise and economical than what it would have taken to describe a structure with no repeatable symmetry at all.

Once again, Theaetetus would have been very surprised to learn of all the applications in which his polyhedra are hidden.

Time to leave the Parc de la Villette and pick up the chronological flow of our history again. How did ancient mathematicians such as Theaetetus come to raise ever more general and theoretical questions? To understand this, we must go back several thousand years to the eastern rim of the Mediterranean.

While the Babylonian and Egyptian cultures were slowly declining, ancient Greece was about to enter its greatest centuries. From the sixth century BC onwards, the Greek world entered an unprecedented period of cultural and scientific effervescence. Philosophy, poetry, sculpture, architecture, theatre, medicine and even history were also disciplines that underwent a true revolution. To this day the exceptional vitality of that period retains a certain fascination and mystery, and last but not least in that vast intellectual movement, mathematics had pride of place.

When we think of ancient Greece, the first image that springs to mind is often that of the city of Athens, dominated by its Acropolis. There, we imagine citizens in white robes, who had just invented the first democracy in history, strolling among Pentelic marble temples and a few olive trees. But this vision comes nowhere near to a representation of the whole of the Greek world in all its diversity.

In the eighth and seventh centuries BC, a swarm of Greek colonies spread around the edges of the Mediterranean. Sometimes there was an integration between these colonies and the local peoples, with a partial adoption of the latter's customs and lifestyles. It is far from the case that all Greeks shared the same form of existence. Their food, pleasures, beliefs and political systems varied widely from one region to another.

Thus, Greek mathematics did not emerge in a specific place where all the scholars knew each other and met each other daily, but over a vast geographic and cultural region. The driving forces behind the mathematical revolution included the contact with the older civilizations whose heir it became, and the mix of Greece's own diversity. Numerous scholars undertook a pilgrimage to Egypt or the Middle East at some stage in their lives, as an obligatory rite of passage in their training. A sizeable amount of Babylonian and Egyptian mathematics was integrated and extended by Greek scholars.

The first great Greek mathematician, Thales, was born at the end of the seventh century in the town of Miletus (present-day Milet), on the south-west coast of what is now Turkey. Although he is mentioned in various sources, it is difficult today to extract reliable information about his life and his work. As with many scholars of this period, various legends spread by overzealous disciples grew up after his death, and have made it hard to unravel the true from the false. The scholars of that time were not averse to spinning yarns, and they not infrequently came to an accommodation with the truth when it did not suit them.

According to one of the many stories put about on the subject, Thales was particularly absent-minded: he is said to have been a prime example from an age-old tradition of scatter-brained scholars. One tale reports that one night he was seen to fall into a well while roaming with his nose in the air observing the stars. Another tells us that he died aged almost eighty while attending a sporting event: so wrapped up did he become in the spectacle that he forgot to eat and drink.

His scientific prowess is also the subject of singular tales. Thales is viewed as the first person to have correctly predicted a solar eclipse. This one took place in the middle of a battle between the Medes and the Lydians, on the banks of the Halys River in the west of what is now Turkey. When night intervened in broad daylight, the warriors saw it as a message from the gods, and decided at once to make peace. Nowadays, the prediction of eclipses or their reconstitution in the past has become child's play for our astronomers. Thanks to them, we know that this eclipse took place on 28 May 584 BC, making the battle of Halys the oldest historical event that we are able to date with such accuracy.

Thales' greatest success was achieved on a journey to Egypt. The story goes that the Pharaoh Amasis in person challenged him to measure the height of the Great Pyramid. Until then, the Egyptian scholars who had been consulted had failed to answer the question correctly. Thales not only met the challenge, but did so in an elegant way, using a particularly clever method. He planted a stick vertically in the ground and waited for the time of day when the length of its shadow was equal to its height. At that very same moment, he marked and then measured the shadow of the pyramid, which likewise must be equal to its height. That did the trick!

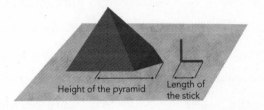

Height of the pyramid   Length of the stick

The story is certainly entertaining, but once again its historical truth is dubious. Moreover, as it is told, the anecdote is quite scornful of the Egyptian scholars of the period, although papyruses such as that copied by Ahmes show that these scholars knew perfectly well how to calculate the height of their pyramids more than a thousand years before Thales came on the scene. So where does the truth lie? Did Thales really measure the height of the pyramid? Was he the first to use the shadow method? And what if he had simply measured the height of an olive tree in front of his house in Miletus, and his disciples took it upon themselves to embroider the story after his death? We have to abide by the evidence, and we shall probably never have it.

Be that as it may, Thales' geometry, on the other hand, is very real, and whether he applied it to the Great Pyramid or to an olive tree, the shadow method is no less appealing. This method represents a particular case of a property now known as Thales' theorem. Several other mathematical results are also attributed to Thales: the circle is divided in two equal parts by any diameter (fig. 1); the angles at the base of an isosceles triangle are equal (fig. 2); the opposite angles produced at the point of intersection of two straight lines are equal (fig. 3); if the three vertices of a triangle lie on a circle and one side passes through the centre of this circle, then the triangle is a right-angled triangle (fig. 4). In fact, this last statement is sometimes also referred to as Thales' theorem.

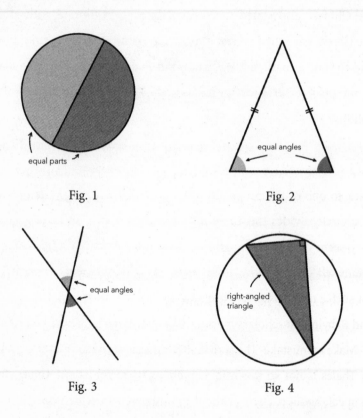

Fig. 1   Fig. 2

Fig. 3   Fig. 4

I want to look at this strange word that is as fascinating as it is awesome: what is a theorem? Etymologically, the word comes from the Greek *thea* (contemplation) and *horáō* (to see), so a theorem would be a sort of observation on the world of mathematics, a fact that would have been noted, examined and then recorded by mathematicians. Theorems can be transmitted orally or in writing, and are similar to one's grand-mother's recipes or to weather lore that has been tested over the generations and is confidently believed to be true. One swallow does not make a summer, bay leaves soothe rheumatism, and the 3-4-5 triangle has a right angle. These are things we believe to be true and that we keep in mind in order to use them when they are relevant.

According to this definition, the Mesopotamians, the Egyptians and the Chinese also stated theorems. However, beginning with Thales, the Greeks gave them a new dimension. For them, a theorem not only had to state a mathematical truth, but that truth must be formulated in the most general way possible, and accompanied by a proof of its validity.

Back to one of the properties attributed to Thales: that the diameter of a circle divides the circle into two equal parts. Such an assertion may seem disappointing coming from a scholar of Thales' calibre. It seems self-evident; why did we have to wait until the sixth century BC for such a trivial assertion to be stated? There is no doubt that Egyptian and Babylonian scholars must have known that a long time beforehand.

Make no mistake. The audacious thing about that property attributed to Thales is not so much its content as its formulation. Thales dared to speak about a circle without saying precisely which one. In order to state the same rule, the Babylonians, Egyptians and Chinese would have used an example. Draw a circle of radius 3 and draw one of its diameters, they would have said, and this circle is divided into two equal parts by this diameter. And when one example did not suffice for an understanding of the rule, they would have given a second one, a third, and a fourth if necessary. They would have given as many examples as needed to enable readers to understand that they could repeat the same procedure on every circle they might meet. But the general assertion was never formulated.

Thales reached a milestone.

'Take a circle, any one you like, I don't need to know which. It may be enormous or tiny. Draw it in the horizontal, in the vertical or on an inclined plane, it doesn't matter to me. I don't care at all about your particular circle and how you have drawn it. However, I assert that its diameter divides it into two equal parts.'

With this operation, Thales definitively assigned the status of abstract mathematical objects to geometric figures. This step in thinking was similar to that taken 2,000 years earlier by the Mesopotamians when they considered numbers independently from the objects counted. A circle was no longer a figure drawn on the ground, on a tablet or on a papyrus. The circle became a fiction, an idea, an abstract ideal all of whose real representations are merely imperfect instances.

From that point on, mathematical truths were stated in a concise and general manner, independently of the various particular cases they covered. It is these statements that the Greeks then called theorems.

Thales had several disciples in Miletus. The two most famous were Anaximenes and Anaximander. Anaximander in turn had his own disciples, who included a certain Pythagoras, who would give his name to the most famous theorem of all time.

Pythagoras was born at the start of the sixth century BC on the island of Samos, situated off the coast of what is now Turkey, and less than 40 kilometres away from the town of Miletus. As a young man he gained experience through his travels in the ancient world, and chose to settle in the town of Croton, in the south-east of present-day Italy. It was there that he founded his school in 532 BC.

Pythagoras and his disciples were not just mathematicians and scholars, but also philosophers, monks and politicians. Yet it must be said that if we were to transpose it to our time, the community begun by Pythagoras would undoubtedly be perceived to be among the most obscure and most dangerous of sects. The life of the Pythagoreans was governed by a set of precise rules. For example, anyone wishing to join the school had to go through a period of five years of silence. The Pythagoreans had no individual possessions: all their belongings were shared. They used various symbols such as the *tetractys* or the pentagram in the shape of a star with five branches to recognize one another. Moreover, the Pythagoreans thought of themselves as enlightened people, and thought it normal that political power should come to them. They firmly repressed the rebellions of towns that refused to accept their authority. In fact, it was in one such riot that Pythagoras died at the age of eighty-five.

The number of myths of all kinds that were invented around Pythagoras is also impressive. His disciples were scarcely lacking in imagination, as we can now see. According to them, Pythagoras was the son of the god Apollo. The name Pythagoras also means literally 'he who was announced by the Oracle': the Oracle of Delphi was in fact the oracle of the temple of Apollo, and it is she who is said to have told Pythagoras' parents about the forthcoming birth of their child. According to the Oracle, Pythagoras would be the most handsome and wisest of men. After such a birth, the Greek scholar was predestined for great things. Pythagoras remembered all his previous lives. According to this, he had in particular been one of the heroes of the Trojan War called Euphorbus. In his youth, Pythagoras took part in the Olympic Games and took the laurels in all the *pygmachia* events (pugilism, the ancestor of our boxing). Pythagoras invented the very first musical scales. Pythagoras was able to walk in the air. Pythagoras died and was resuscitated. Pythagoras was a talented soothsayer and healer. Pythagoras had control over animals. Pythagoras had a golden thigh.

While most of these legends are so far-fetched that no one believes them, in other cases the jury is still out. Is it true, for example, that Pythagoras was the first to use the word 'mathematics'? The facts are so sketchy that some historians have even come to speculate that Pythagoras was a purely fictional person, dreamed up by the Pythagoreans to serve as their tutelary figure.

Therefore, since it is not possible to learn more about the man, let us return to the subject for which he is still known to schoolchildren the whole world over more than 2,500 years after his death: Pythagoras' theorem. What does this famous theorem tell us? Its statement may seem astonishing, because it establishes a link between two apparently unconnected mathematical concepts: right-angled triangles and the square numbers.

Let us return to our favourite triangle, the 3-4-5. From the lengths of the three sides we can construct three square numbers: 9, 16 and 25.

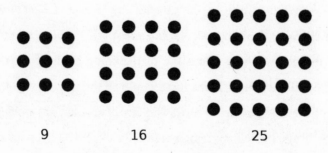

9   16   25

We now spot a curious coincidence: $9 + 16 = 25$. The sum of the squares of sides 3 and 4 is equal to the square of side 5. This might be put down to chance, yet if we try to reproduce this calculation for a different right-angled triangle, it still works. Consider, for example, the triangle 65-72-97, which appears on the Babylonian Plimpton tablet. The three corresponding squares are 4,225, 5,184 and 9,409. And Pythagoras' theorem does not fail: $4,225 + 5,184 = 9,409$. For such large numbers, it becomes difficult to believe in a simple coincidence.

You can try with all the right-angled triangles you like, small or large, fat or thin, it always works. In a right-angled triangle, the sum of the squares of the two sides that form the right angle is always equal to the square of the third side (which we call the hypotenuse). And it also works the other way round: if, in a given triangle, the sum of the squares of the two smallest sides is equal to the square of the largest side, then the triangle is right-angled. That is Pythagoras' Theorem.

Of course, we do not know for certain whether Pythagoras or his disciples actually contributed to this theorem. Even though the Babylonians never expressed it in the general form that we have just seen, it is highly likely that they already knew this result more than one thousand years earlier. For otherwise, how would they have been able to discover all the right-angled triangles that appear on the Plimpton tablet with such accuracy? The Egyptians and the Chinese probably also knew the theorem, which was also clearly stated in the commentaries that were added to the *Nine Chapters* in the centuries after it was written.

Some accounts claim that Pythagoras was the first to give a proof of the theorem. However, there is no reliable source to confirm this, and the oldest proof that has come down to us is only the one in Euclid's *Elements*, which was written three centuries later.

5

A LITTLE METHOD

The matter of the proof became one of the main fields of work for Greek mathematics. No theorem could be validated without being accompanied by a proof, meaning a piece of precise logical reasoning establishing its truth in a definitive manner. It has to be said that without the safeguard that proofs provide, mathematical results can sometimes hold some nasty surprises. Certain methods do not always work as well as they might, even though they may be recognized and widely used.

Do you recall the construction of the Rhind Papyrus for drawing a square and a circle of the same area? Well, it is incorrect. Only a little, certainly, but incorrect all the same. Measure the areas precisely, and they differ by about 0.5 per cent. While I agree that for surveyors and other geometers of land such accuracy may broadly suffice, it won't do for theoretical mathematicians.

Pythagoras himself was caught out by false hypotheses. His most famous mistake concerned commensurable lengths. He believed that in geometry, two lengths are always commensurable – in other words, that it is possible to find a unit small enough that it will measure them both simultaneously. Imagine a line of length 9 centimetres and another of length 13.7 centimetres. The Greeks did not know about numbers with decimal points, they only measured lengths with whole numbers. Thus, for them, the second line was not measurable in centimetres. Never mind that, in this case, it is sufficient to take a unit ten times smaller to say that the two lines measure respectively 90 and 137 in millimetres. Pythagoras had convinced himself that regardless of their lengths, any two lines were always commensurable for some appropriate unit of measurement which remained to be determined.

However, his conviction was disproved by a Pythagorean called Hippasus of Metapontum. He discovered that in a square, the side and the diagonal are not commensurable. Whatever unit of measurement you may choose, it is not possible to measure both the side of the square and its diagonal simultaneously and arrive at a result in whole numbers. We call numbers like this 'irrational'. Just as with the side of the square and its diagonal, it is equally impossible to measure the diameter and circumference of a circle in commensurable numbers. Hippasus provided a logical proof of this, which left no room for doubt about the question. Pythagoras and his disciples were so annoyed by this that Hippasus was excluded from the School. His discovery led to his being taken out to sea and thrown overboard by his fellow disciples – or so the story goes.

These anecdotes are terrifying for mathematicians. Can we never be sure about anything? Must we live in permanent fear that every mathematical discovery will eventually collapse? And what about the

3-4-5 triangle? Can we be absolutely certain that it is right-angled? Is there not a risk of one day discovering that the angle that, until then, seemed to be a perfect right angle is also only approximate?

It is still not uncommon for mathematicians to fall victim to deceptive intuition. That is why, following the search for rigour by their Greek counterparts, our mathematicians now take great care to differentiate between proven statements that they call 'theorems' and those known as 'conjectures', which they believe to be true, but for which they do not yet have a proof.

One of the most famous conjectures of our time is the so-called Riemann hypothesis. Numerous mathematicians are confident enough that this unproven hypothesis is true that they incorporate it as the basis for their research. The hope is that this conjecture will one day become a theorem, and all their work will then be validated. But if it were to be contradicted one day, all these bodies of work representing whole lifetimes of research would collapse with it. Scientists of the twenty-first century may be more reasonable than their Greek ancestors. However, we can understand that under these conditions, the mathematician announcing the invalidity of the Riemann hypothesis might make some of his colleagues want to throw him to the sharks.

It is in order to escape this permanent unease over refutation that mathematics needs proofs. No, we shall never discover that the 3-4-5 is not a right-angled triangle. It is one – that is for certain. And this certainty comes from the fact that Pythagoras' theorem has a proof. Every triangle for which the sum of the squares of the two sides is equal to the square of the third side is a right-angled triangle. This statement was undoubtedly only a conjecture for the Mesopotamians. It became a theorem with the Greeks. Phew.

But then, what does a proof look like? Pythagoras' theorem is not just the most famous of all theorems, it is also among those that have the greatest number of different proofs. There are at least a few dozen. Some of these were discovered independently by civilizations that had never heard of Euclid or of Pythagoras, as happens, for example, with the proofs that we find in the commentaries on the Chinese *Nine Chapters*. Others are the work of mathematicians who knew that the theorem was already proved but who, as a challenge or in order to leave their personal stamp on it, had fun establishing new proofs. The latter include a few famous names such as the Italian Leonardo da Vinci and the twentieth President of the United States of America, James Abram Garfield.

One principle that we find in several of these proofs is that of the jigsaw puzzle: if two geometric figures can be formed from the same pieces, then they have the same area. Look at this breakdown conceived by the third-century Chinese mathematician Liu Hui.

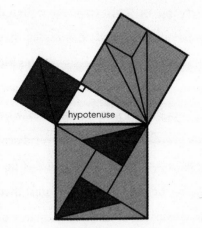

The two squares constructed on the two sides forming the right angle of the right-angled triangle consist of two and five pieces, respectively. These same seven pieces also go to form the square constructed on the hypotenuse. The area of the square on the hypotenuse is thus equal to the sum of the areas of the two smaller squares. And since the area of a square is equal to the square number associated with the length of its side, this shows that Pythagoras' theorem is indeed true.

We shall not go into the details here, but of course, for the proof to be complete we need to show that all the pieces are rigorously identical and that such a breakdown works for all right-angled triangles.

To summarize, let us return to the chain of our deductions. Why is the 3-4-5 a right-angled triangle? This is because it satisfies Pythagoras' theorem. And why is Pythagoras' theorem true? This is because Liu Hui's decomposition shows that the square on the hypotenuse is formed from the same pieces as the two squares on the sides forming the right angle. This has the air of the 'Why?' game that children love to play. The problem is that this little game has the annoying defect of being never-ending. Whatever answer is given to a question, it is always possible to question this answer in turn – 'Why?' Yes, why?

Back to our jigsaw: we have asserted that if figures are constructed from the same pieces, they have the same area. But have we proved that this principle is always true? Might one not be able to find jigsaw pieces whose area varies according to how they are assembled?

Such a proposition seems absurd, doesn't it? It might seem so absurd that it would be eccentric to try to prove it. However, we have just recognized that it is important to prove everything in mathematics. Would we be prepared to give up our principles, just a few moments after having adopting them?

The situation is serious. In fact, even if we were to succeed in explaining why the jigsaw principle is true, we would still have to justify the reasoning we used to that end.

The Greek mathematicians were well aware of this problem. To produce a proof, you have to begin somewhere. Though the very first phrase of any work in mathematics cannot have been proved, precisely because it is the first. Every mathematical construction must therefore begin by admitting a certain number of preliminary facts. These facts will form the foundations for all the deductions that will follow, and will thus need to be chosen with the greatest possible care.

Mathematicians call these facts 'axioms'. Axioms are mathematical statements, just as theorems and conjectures may be, the difference being that unlike the latter, they do not have a proof and do not seek to have one. They are admitted as being true.

Euclid's *Elements*, written in the third century BC, forms a set of thirteen books dealing mainly with geometry and arithmetic.

Little is known about Euclid, and the sources relating to him are very much rarer than for Thales or Pythagoras. He may have lived around Alexandria. Others have suggested, as had already been mentioned for Pythagoras, that this may not have been one man, but the collective name of various scholars. Nothing can be taken for granted.

Despite the small amount of information that we have about him, Euclid bequeathed a monumental work. The *Elements* is unanimously considered to be one of the greatest texts in the history of mathematics for having been the first to adopt an axiomatic approach. The work's construction is surprisingly modern, and its structure is very similar to that still used by the mathematicians of our day. At the end of the fifteenth century, the *Elements* was among the very first works to be printed on the new Gutenberg presses. Euclid's work is today the second-most published text in history, just behind the Bible.

In the first book of the *Elements*, which deals with plane geometry, Euclid introduces the following five axioms or postulates:

1. **A STRAIGHT LINE CAN BE DRAWN FROM ANY POINT TO ANY POINT;**

2. **A FINITE STRAIGHT LINE SEGMENT CAN BE EXTENDED CONTINUOUSLY IN A STRAIGHT LINE ON BOTH SIDES;**

3. **GIVEN A FINITE STRAIGHT LINE SEGMENT, ONE CAN DRAW A CIRCLE WHOSE RADIUS IS THIS SEGMENT AND WHOSE CENTRE IS ONE END OF THIS SEGMENT;**

4. **ALL RIGHT ANGLES ARE EQUAL TO ONE ANOTHER;**

5. **IF A STRAIGHT LINE FALLING ON TWO STRAIGHT LINES MAKES INTERIOR ANGLES ON THE SAME SIDE WHOSE SUM IS LESS THAN TWO RIGHT ANGLES THEN THE TWO STRAIGHT LINES IF EXTENDED CONTINUOUSLY MEET ON THAT SAME SIDE.***

There then follows a host of impeccably well-proved theorems. For each of these, Euclid uses only the above five axioms or the results that he has previously established. The very last theorem of the first book was old knowledge, since it is Pythagoras' theorem.

After Euclid, many mathematicians investigated in their turn the question of the choice of axioms. Many were particularly intrigued

---

* This axiom, which is clearly more complex than the four others, led to numerous debates among mathematicians. In the figure below, the sum of the two angles indicated is less than two right angles; consequently, lines 1 and 2 intersect on the same side as these two angles.

and disturbed by the fifth axiom. This axiom is actually much less elementary than its four associates. It is sometimes replaced by another simpler statement, which allows us to reach the same conclusions: *Given a point, there is precisely one line that can be drawn through that point parallel to a given line.* The debates on the choice of the fifth axiom persisted until the nineteenth century, when they came to a head in the creation of new geometric models in which this axiom is false.

The statement of the axioms poses another problem: that of definitions. What do all these words that are used here mean: points, straight lines, angles, circles? Just as for proofs, the question of definitions is endless. The first definition introduced has to be properly expressed with words that will not have been previously defined.

In the *Elements*, the definitions precede the axioms. The first phrase of the first book is the definition of a point.

A point is that which has no part.

Figure that out! What Euclid wished to say in this definition is that the point is the smallest possible geometric figure. It is impossible to make jigsaws out of a point, it cannot be cut up, it has no parts. In 1632, in one of the first French editions of the *Elements*, the mathematician Denis Henrion fills out the definition a little in his commentaries by stating that a point has no length, no breadth and no thickness.

These negative definitions make us sceptical. To say what a point is not, is not exactly to say what it is. It would take someone clever to propose something better. In some early twentieth-century reference books you sometimes found the following definition: *A point is the trace left by a finely sharpened pencil when one presses on a sheet of paper*. 'Finely sharpened'! This time, the definition is concrete. However this definition would have seriously exercised Euclid, Pythagoras and Thales, who had taken so much trouble to produce geometric figures out of abstract and idealized objects. No pencil, however finely sharpened it might be, could leave a trace that truly had no length, no breadth and no thickness.

In short, no one really knows how to say what a point is, but everyone is more or less convinced that the idea is sufficiently simple and clear that it cannot lead to ambiguity. We are all more or less certain that we are talking about the same thing when we use the word *point*.

It was based on this leap of faith in the first definitions and in the axioms that the whole edifice of geometry was built. And, for lack of anything better, this same model forms the basis for the construction of all modern mathematics.

DEFINITIONS – AXIOMS – THEOREMS – PROOFS

The path Euclid followed set the pattern of the mathematics that came after him. However, while the theories became structured and expanded, new grains of sand worked their way into the mathematicians' shoes: paradoxes.

One definition of a paradox is that it is something that should work, but does not work. It is an apparently insoluble contradiction, an argument that seems perfectly correct but ends in a completely absurd result. Just imagine if you were to set up a list of axioms that seemed obvious to you, and that you were able to deduce from them theorems that were clearly false. What a nightmare that would be.

One of the most famous paradoxes is attributed to Eubulides of Miletus, and concerns statements made by the poet Epimenides. The latter is said to have declared one day that 'Cretans are liars'. But Epimenides was himself a Cretan. Consequently, if what he said was true, he was a liar, and so what he said was false. And if, conversely his declaration was false, then he was lying and the declaration was telling the truth. Several versions of the same paradox have subsequently been invented, where the simplest of them simply involves someone saying: 'I am lying.'

The liar's paradox calls into question a preconceived idea according to which every sentence must be either true or false. There is no third possibility. In mathematics, this is known as the principle of the excluded third. At first sight, it would be quite tempting to make this principle an axiom. However, the liar's paradox puts us on our guard: the situation is more complex than that. If a statement asserts its own falsehood, then logically it cannot be either true or false.

This curiosity has not prevented the majority of mathematicians even up to now from considering the excluded third to be true. After all, the liar's paradox is not really a mathematical statement, and one could view it more as a linguistic inconsistency than a logical contradiction. However, more than two thousand years after Eubulides, logicians discovered that paradoxes of the same type can also appear within even the most rigorous of theories, leading to a profound upheaval in mathematics.

Zeno of Elea, a Greek who lived in the fifth century BC, was also a past master of the art of creating paradoxes. Some ten paradoxes are attributed to him. One of the most famous ones is 'Achilles and the Tortoise'.

Imagine a race between Achilles, who was a remarkable athlete, and a tortoise. In order to even out the chances, the tortoise is given a head start of, say, 100 metres. Despite this start, it seems to be accepted that Achilles, who runs much faster than the tortoise, will sooner or later catch up with the latter. However, Zeno asserted the opposite.

'Let us study the race in several stages,' he said. In order to be able to catch up with the tortoise, Achilles has to run at least the 100 metres separating him from the latter. In the time he takes to cross this distance, the tortoise will also have advanced a little and so Achilles will still have to cross a certain distance in order to catch up with it. But when he has crossed that distance, the tortoise will have advanced a little more. Achilles will then still have to cross another short distance after which the tortoise will again have advanced a little.

In short, every time Achilles reaches the point previously occupied by the tortoise, the latter has moved on a little and is still not caught. And this remains true however many stages one considers. Achilles thus seems to be doomed to become ever closer to the tortoise without ever being able to overtake it.

This is absurd isn't it? You only have to try this out to see that the runner will definitely end up by overtaking the tortoise. And yet, the reasoning seems to stand up, and it appears difficult to discern a logical error in it.

It took mathematicians a long time to understand this paradox, which cleverly disregards the infinite. If the runners go in a straight line, their trajectory can be compared with what Euclid calls a segment. A segment has a finite length even though it consists of infinitely many points all of length zero. Thus, in a certain way, there is something infinite in the finite. Zeno's paradox subdivides the time interval

Achilles will take to catch the tortoise into infinitely many increasingly small intervals.

However, this infinite number of steps actually occurs in a finite time, and that in no way prevents Achilles from catching the tortoise once this time elapses.

The notion of infinity in mathematics is undoubtedly the greatest source of paradoxes, but it is also the cradle of the most fascinating theories.

Throughout history, mathematicians have maintained an ambiguous relationship with paradoxes. On the one hand paradoxes represent the greatest threat to them. The danger is that one day a theory might produce a paradox that would result in the collapse of all its foundations, and hence of all the theorems that we thought we had built up upon its axioms. But on the other hand, these would be amazing challenges. Paradoxes are a very prolific and stimulating source of questioning. If there is a paradox, then something must have escaped us. We must have misunderstood a concept, introduced a poor definition or chosen an axiom poorly. We must have been accepting something as a fact when it is not. Paradoxes are an invitation to adventure. They invite us to rethink everything up to and including our most intimate facts. How many new ideas and original theories would we have passed by, if paradoxes had not been there to push us towards them?

Zeno's paradoxes inspired new notions about infinity and measure. The liar's paradox led logicians into a search for ever-sharper notions about truth and provability. To this day, numerous researchers are still dissecting mathematics whose seeds were already present in the paradoxes of Greek scholars.

In 1924, the mathematicians Stefan Banach and Alfred Tarski revealed a paradox that now carries their names and that calls into question the very principle of jigsaw puzzles. However evident it may appear, this principle may be at fault. Banach and Tarski were able to describe a jigsaw puzzle in three dimensions in which the volume varies according to how the fragments are fitted together. We shall return to this. The pieces they imagined are, however, so strange and weird-looking that they have nothing to do with the geometric figures that the Greek geometers dealt with. Rest assured, the jigsaw principle remains valid as long as the pieces are shaped like triangles, squares or other classical figures. Liu Hui's proof of Pythagoras' theorem still holds.

But let that be a lesson to us: we should be suspicious of facts and let ourselves be filled with wonder and surprise by the mysteries of this world of mathematics that the Greek scholars opened up to us.

6

π IN THE SKY

It is 14 March 2015, I am visiting the Palais de la Découverte ('Palace of Discovery'), the science museum in the Grand Palais des Champs-Élysées in Paris, where there is a celebration today.

At the start of the 1930s, the French physicist and Nobel Prize Laureate Jean Perrin conceived a project to establish a centre for science to increase the public's interest in advances in research in all areas of science. The Palais de la Découverte was created in 1937, just a few steps off the Champs-Élysées, where it took over the whole of the west wing of the Grand Palais, with an area of 25,000 square metres. The exhibitions, which were only meant to last six months, were so successful that from 1938 onwards, what was temporary became permanent. Eighty years after it opened, the venue still receives several hundred thousand visitors per year.

On leaving the Métro, I go up the Avenue Franklin-Roosevelt towards the entrance to the Palace. I arrive at the front steps, and there, a detail attracts my attention: *4, 2, 0, 1, 9, 8, 9*. A strange procession of printed figures snakes its way across the ground, climbs the steps and seems to thread its way into the interior of the building. How unusual.

The last time I came this way, they weren't there. I follow them: *1, 3, 0, 0, 1, 9*. I enter the Palace. They're still there, *1, 7, 1, 2, 2, 6*. They cross the central rotunda and head off towards the grand staircase, *7, 6, 6, 9, 1, 4*. I bound up the steps four at a time, pass in front of the entrance to the planetarium and turn to the left, *5, 0, 2, 4, 4, 5*. The figures are leading me straight on to the mathematics department. I see them curl like ivy, leaving the ground and climbing along the wall, *5, 1, 8, 7, 0, 7*. Finally they return to their source. I am in the middle of a wide circular room, the black and red numbers have grown, and they whirl around getting higher and higher. Finally, my eyes land on the start of the sequence: *3, 1, 4, 1, 5* . . . I am in the heart of one of the most symbolic rooms of the Palais de la Découverte: the π Room.

The number π ('pi') is undoubtedly the most famous and most fascinating of mathematical constants. The circular shape of the room reminds me that its value is intimately linked to the geometry of the circle: this is the number by which you have to multiply the diameter of a circle to find its perimeter. The letter π is in fact the sixteenth letter of the Greek alphabet, equivalent to our 'p', and the first letter of the word perimeter. The number π is not very large, scarcely more than 3, but its decimal expansion is infinite: 3.14159265358979 . . .

Usually, visitors can only see the first 704 decimal digits of the number snaking around the circular walls of the π Room, but today the numbers are out and about. They are invading the whole of the Palace and are on show right out to the street. There are now more than 1,000 decimal digits. I should say that it is an historic date: 14 March 2015 is the π Day of the century!

The first π Day was launched on 14 March 1988 at the Exploratorium, the American cousin of the Palais de la Découverte, located in the heart of San Francisco. The fourteenth day of the third month, or 3/14, was the perfect date for celebrating π, where 3.14 is the usual approximation to two figures after the decimal point. Since then, the initiative has been emulated, and numerous fans around the world come together every year to celebrate the constant and, through it, mathematics. The celebration took off to such an extent that in 2009, the 'π Day' was officially recognized by the United States House of Representatives.

This year, the aficionados of π awaited their day with even greater impatience. Today's date is 3/14/15, and an extra two figures add to the coincidence of the date and the constant. This version has to be magnificent. On this occasion the whole of the mathematics team of the Palais de la Découverte is on the bridge. This is also the reason why I am here. Together with a few other maths buffs, we have come to lend a hand for a day that promises to be rich in mathematical experiences.

While the number π revealed itself through geometry, it then spread into most branches of mathematics. It is a number with many faces. In arithmetic, in algebra, in analysis, in probability, mathematicians, of whatever discipline, who have never had anything to do with π are rare. In the very heart of the Palais de la Découverte, the rotunda is teeming with activities that show its many facets. Here, visitors are invited to count needles thrown at random onto a floor tile. There they are observing the proportion of numbers appearing in multiplication tables. On the floor, children are covering the surface of a disc with wooden boards. Another group is busy studying the trajectory of a fixed point on a wheel rolling on a plane. And everyone ends up finding the same result: 3.1415 . . .

Further away, a computer program invites visitors to look for the first occurrences of their date of birth in the sequence of decimals. A young man born on 25 September 1994 tries it. The result comes up almost immediately: the sequence 25091994 occurs in the number π starting at the 12,785,022th decimal digit. Mathematicians have conjectured that all sequences of numbers, however long they may be, occur at one point or another in the decimal expansion of π. Computer simulations appear to confirm this: up to now, all the sequences that have been investigated have been eventually found. However, no one has yet been able to give the incontestable proof that this will always be the case.

A twelve-year-old girl comes up to me. She seems intrigued by the strange instruments that surround us, and gives me an inquisitive look.

'You're wondering what all that is about, aren't you? Have you heard about the number π?'

'Oh yes! she says. It's 3.14. Well, no . . . It's almost 3.14 . . . We saw that at school. It's for calculating the perimeter of a circle. We learned some poetry too.'

'Poetry?'

Her eyes narrow as though it helps her dig into her memory, then she begins to recite.

QUE J'AIME À FAIRE APPRENDRE CE NOMBRE UTILE AUX SAGES

IMMORTEL ARCHIMÈDE, ARTISTE, INGÉNIEUR,

QUI DE TON JUGEMENT PEUT PRISER LA VALEUR?

POUR MOI TON PROBLÈME EUT DE PAREILS AVANTAGES.

I smile while listening to this rhyme, which I had also learned when I was her age. I had forgotten it. The way it works is particularly ingenious: to reconstitute the start of the number π, you just have to count the number of letters in each word. '*Que*' = 3; '*j*' = 1; '*aime*' = 4; and so on. There are numerous versions of this poem in different languages. 'Near a Raven', one of the most famous versions, is an adaptation of 'The Raven' by Edgar Allan Poe. It can be used to find 740 decimal digits* and follows a similar process: 'Poe' = 3; 'E' = 1; 'Near' = 4; and so on.

'Well done!' I say. 'I don't think I would have remembered it quite as well. But tell me, you just spoke about Archimedes then, in your poem? Do you know who he was?'

I'm testing her. She pouts, then shrugs.

It's time for a journey; time for a refresher. I unfold a large articulated circle, subdivided into a multitude of interlocking triangles. We're flying off to the ancient Sicilian town of Syracuse and 2,300 years back. That's where Archimedes is waiting for us.

The cicadas are singing under a leaden sky. The streets are filled with perfumes from the four corners of the Mediterranean. Olives, fish and grapes lie side by side on the market stalls. To the north of the town, the imposing silhouette of Mount Etna looms on the horizon. To the west, the fertile plains ensure that the colony prospers, while to the east the double harbour opens out to the sea. Syracuse gained

---

* Edgar Allan Poe's poem 'The Raven', written in 1845, was adapted in 1995 by Michael Keith under the title 'Near a Raven' as a mnemonic for the mathematical constant. It starts: '*Poe, E., / Near a Raven. / Midnights so dreary, tired and weary. / Silently pondering volumes extolling all by-now obsolete lore. / During my rather long nap – the weirdest tap! / An ominous vibrating sound disturbing my chamber's antedoor. /"This", I whispered quietly, "I ignore."*' . . .'

its fame and its power by establishing itself as one of the key maritime crossroads of the region. Founded five centuries earlier by Greek colonists from Corinth, the town was one of the most prosperous in the Mediterranean region.

It is there that in 287 BC a man whose genius and inventiveness would inaugurate a new style of mathematics was born. Archimedes ranks among the great inventors, the problem solvers, those capable of having resolutely new and revolutionary ideas. We owe the principle of the lever and also that of the screw to him. According to legend, it was he who let out the famous cry '*Eureka!*' – 'I've found it!' – one day when he was in his bath; he had just hit upon the physical principle that now carries his name: any body immersed in a liquid is subject to an upwards force with an intensity equal to the weight of the liquid displaced. This is why objects lighter than water float, while those that are heavier sink. The story also goes that at a time when Syracuse was besieged by the Roman fleet, Archimedes invented a system of mirrors to focus the sun's rays so as to set fire to the enemy vessels.

In mathematics, it is to Archimedes that we owe the first great advances on the trail of the number π.

Others before him had been interested in the circle, but their approaches often lacked rigour. Remember the Chinese *Nine Chapters*: there we find circular fields of diameter 10 *bu* with a circumference of 30 *bu*. Such information amounted to saying that the number π was equal to 3. In Ahmes' papyrus, the approximate solution for the squaring of the circle was equivalent to considering a value for π of about 3.16.

Archimedes knew that it was difficult, or literally impossible, to calculate an exact value for π, so he too had to make do with approximations, but his approach had two distinctive features. First, where his prede-cessors thought they might find an exact method, Archimedes was perfectly aware that only approximate values could ever be achieved. Second, he estimated the difference between his approximations and the true value of π, then developed techniques to reduce this discrep-ancy further and further.

Based on calculations, he finally concluded that the value he was looking for lay between two numbers which, written in our present-day decimal system, were approximately equal to 3.1408 and 3.1428. In short, Archimedes now knew the number π to within 0.03 per cent.

ARCHIMEDES' METHOD

To calculate his window for π, Archimedes approximated the circle by regular polygons. Let us take, for example, a circle whose diameter measures one unit and whose circumference therefore measures π units, and then frame it in a square.

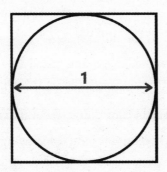

The square has a side equal to 1 (like the diameter of the circle) and thus a perimeter equal to 4. Since the circumference of the circle is smaller than the perimeter of the square, it follows that π is less than 4.

On the other hand, we can inscribe a hexagon in the circle, as here:

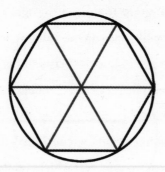

The hexagon consists of six equilateral triangles whose sides are of length 0.5 units (half of the diameter of the circle). The perimeter of the hexagon is thus of length 6 × 0.5 = 3 units. It follows that π is greater than 3!

Nothing exciting so far, the framing between 3 and 4 is still very imprecise. In order to tighten the window, we can now increase the number of sides of the polygons. If we divide each side of the hexagon into two, we now obtain a figure with 12 sides – a dodecagon – which comes much closer to the circle.

Several intricate geometric calculations later (mainly based on Pythagoras' theorem), we reach the conclusion that the perimeter of the dodecagon has a length of about 3.11. The number π is thus greater than this value.

To obtain his bounds to within 0.001, Archimedes simply repeated this operation three more times. By dividing each side into two, we find polygons with 24, then 48, and eventually 96 sides.

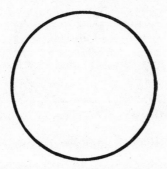

You can't see the polygon? That's to be expected: the sides now adhere so closely to the circle that it becomes almost impossible to distinguish them with the naked eye. This is how Archimedes reached the conclusion that π is greater than 3.1408. And by resuming this process for the polygons outside the circle, he found that π is less than 3.1428.

The power of Archimedes' method lay not only in its result, but also in the fact that it could be extended. By continuing to subdivide our polygons we can refine our bounds again and again. In theory, it is therefore possible to obtain an approximation for the number π that is as precise as you wish, provided you dare to tackle the calculations.

In 212 BC Roman troops finally succeeded in entering Syracuse. The general Marcus Claudius Marcellus, who had led the siege of the town, ordered his soldiers to spare Archimedes, who was then aged seventy-five. However, while his town was in the midst of falling, the Greek scholar was wrapped up in his study of a geometric problem. When a Roman soldier happened to pass him by, Archimedes, who had drawn his figures on the ground, absent-mindedly said to him:

'Don't touch my circles!' The soldier was angry and ran him through with his sword.

General Marcellus built a sublime tomb for him, surmounted by a sphere inscribed in a cylinder, illustrating one of his most remarkable theorems. Over the course of the following seven centuries, the Roman Empire never produced another mathematician of the calibre of Archimedes.

As far as mathematics was concerned, antiquity ended feebly. Soon the Roman Empire extended over the whole of the Mediterranean, and the Greek identity was diluted in this new culture. But one town continued to keep the spirit of the Greek mathematicians alive for several more centuries, and that was Alexandria.

In the course of his conquests, Alexander the Great captured Egypt at the end of the year 332 BC. He only stayed there for a few months, long enough to have himself proclaimed Pharaoh at Memphis and to decide to found a new town on the Mediterranean coast. He never saw the town he gave his name to. When he died eight years later in Babylon, his kingdom was shared between his various generals, and Egypt returned to Ptolemy I Soter, who made Alexandria his capital. Under his reign, Alexandria became one of the most prosperous cities of the Mediterranean basin.

Ptolemy continued the major development works initiated by Alexander. On the tip of the island of Pharos, which faces the town, he embarked on the building of a monumental lighthouse. It did not take Greek authors long to recognize the lighthouse of Alexandria as an absolutely exceptional monument and include it as the seventh and final member of the exclusive list of what we now call the Wonders of the Ancient World.

Let us pause for a few moments to take advantage of the extra-ordinary panorama that can be seen by travellers who have the courage to climb the hundreds of steps of the spiral staircase leading to its top. Look to the north. The Mediterranean Sea stretches away as far as the eye can see. From here, you can spot the incoming merchant ships as they come into view more than 50 kilometres away. There is another one passing in front of you now and entering the harbour, its hold laden with merchandise. It may have come from Athens, from Syracuse or even from Massalia, that dynamic city in the south of Gaul that will one day be called Marseille. Turn to look southward, and you have a view of the Nile Delta. Five kilometres from there, you can make out an expanse of saltwater that crosses the delta: this is Lake Mareotis. Between the lake and the sea, on that broad strip of land, the town of Alexandria lies resplendent. It is a modern new town. Here and there, you can still see building sites.

The lighthouse is not alone on the island of Pharos, which also houses the Temple of Isis. In order to reach it, Alexandrians have to take the Heptastadion, a causeway some 1,300 metres long, which splits the harbour into two separate basins. From the top of the light-house, you can make out the tiny outlines of people walking along it. Those returning to the mainland arrive in the royal quarter where Ptolemy's palace, the theatre and the Temple of Poseidon are located. Slightly further westward, lies an immensely significant building: the Musaeum, or Mouseion.

Ptolemy wished to make Alexandria a great cultural centre capable of rivalling Athens, so he invested in it through this great museum, designed to preserve the Greek cultural heritage. Scholars who came to visit the Mouseion were pampered. They were housed, fed and paid to carry out their work. The king also provided a gigantic library for them, the legendary Library of Alexandria. The fame and prestige of the Mouseion perhaps owed more to that library than to the scientists who worked there.

Ptolemy's strategy for filling the library was simple: all the ships that stopped at Alexandria had to hand over the books that they were transporting. The books were then copied and the copies returned to the ship. As for the original, it went straight into the library's collections. Later, Ptolemy II Philadelphus, the son and successor to Ptolemy I, appealed to all the kings of the world to send him examples of the most famous works of their region. When it opened, the Library of Alexandria already had almost 400,000 volumes. Subsequently, this number rose to as many as 700,000.

Ptolemy's plan worked a treat, and over a period of more than seven centuries, a succession of scholars attended Alexandria, where the intellectual environment preserved the vitality that was in short supply in the rest of the Mediterranean world.

The most famous residents of the Mouseion included Eratosthenes of Cyrene, who was, you will recall, the first to measure the Earth's circumference accurately. It was also there that Euclid is said to have written the majority of his *Elements*. A mathematician named Diophantus wrote a famous work there about the equations that now carry his name. In the second century AD, it was also in Alexandria that Claudius Ptolemy (no relation to Ptolemy II) wrote the *Almagest*, a work that collected a large amount of knowledge about the astronomy and mathematics of his period. Even though Ptolemy stated in it that the sun revolved around the Earth, the *Almagest* remained a reference work until Copernicus arrived to add his contribution in the sixteenth century.

Alexandria was not just home to scholars who wrote down or produced new knowledge. A whole ecosystem of copyists, translators, literary commentators and editors grew up around the Mouseion. The town teemed with all these people.

Unfortunately, the fourth century saw the start of troubled times. On 16 June 391 the Emperor Theodosius I, who wished to speed up the conversion of the Empire to Christianity, issued an edict banning all pagan cults. The Mouseion, although it was not really a temple, was affected by the Emperor's decision and closed in its wake.

At that time, one of the figures of the Alexandrian intellectual milieu was Hypatia. Her father, Theon, was the head of the Mouseion when it was closed. However, that did not prevent the scholars of the town from continuing their work for a time. Socrates Scholasticus later wrote that an almost infinite crowd of people would throng to hear Hypatia speak. She was both a mathematician and a philosopher, and surpassed all the men of her day in her science. She is also the first woman in our story.

Was she really the first? That is not completely true. Other women did mathematics before Hypatia, but their works or their biographies have not been passed down to us. In particular, women were admitted to the School of Pythagoras. The names of several of them, such as Theano, Autocharidas and Habroteleia, are known to us, but it has to be said that we know almost nothing about them.

No texts written by Hypatia have come down to us, but several sources mention her works. She was mainly interested in arithmetic, geometry and astronomy. In particular, she continued the work undertaken several centuries earlier by Diophantus and Ptolemy. Hypatia was also a prolific inventor. Her inventions included the hydrometer, which was used to measure the density of a liquid based on skilful application of Archimedes' principle, together with a new astrolabe model, which facilitated astronomic measurements.

Unfortunately, her story was a short one. In 415, she managed to offend the Christians of the town, who pursued her and finally murdered her. They hacked her body to pieces, and then burnt it.

Following the closure of the Mouseion and the death of Hypatia, Alexandria's scientific flame petered out. The library's collections were not spared. The town was hit by fires, pillage, tidal waves and earthquakes, and although we do not know exactly how and when the library of Alexandria disappeared, the evidence is there: in the seventh century, nothing at all remained of it.

This was the end of an era. But history has many roundabout routes, and Greek mathematics soon found other ways to reach us.

7

NOTHING AND LESS THAN NOTHING

In Tibet, Mount Kailash, with its summit at an altitude of 6,714 metres, forms part of a closed circle of peaks that no human being has ever climbed. Its rounded silhouette, with striations of snow on the grey granite, stands out impressively above the jagged landscape of the Western Himalayas. For the inhabitants of the region, be they Hindus or Buddhists, the mountain is sacred and carries its share of ancestral myths and supernatural stories. It is even said that it is the legendary Mount Meru, which according to the local mythologies marks the centre of the universe.

The source of one of the seven sacred rivers of the region, the Indus, is hidden here. On leaving the slopes of Mount Kailash, the Indus heads eastward, and rapidly zigzags its way through the mountains of Kashmir before it starts to descend slowly southwards. There it crosses the plains of the Punjab and Sind, in what is now Pakistan, before it forms a delta into the Arabian Sea. The Indus Valley is fertile. In antiquity, the region was covered by deep rustling forests. There, Asian elephants could be found along with rhinoceroses, Bengal tigers, monkeys aplenty and snakes, which the snake charmers would later attempt to charm with their flutes. Off the beaten track, one could almost expect to come across Rudyard Kipling's young hero Mowgli, whose adventures took place in such settings. Here, an original and discrete civilization was born, one whose mathematics would play a key role at the start of the Middle Ages.

From the third millennium BC a number of major towns, such as Mohenjo-daro and Harappa, grew around the river. From a distance, these towns built with clay bricks looked a little like their contemporaries in Mesopotamia. The second millennium saw the start of the Vedic period. The region was fragmented into a multitude of small kingdoms, which spread right up to the banks of the Ganges. Hinduism was born, and developed, and the first great texts in Sanskrit were written.

In the fourth century BC Alexander the Great reached the banks of the Indus and founded a town that he named Alexandria, though later inhabitants may not have known of the prestigious destiny of its Egyptian sister. A part of Greek culture was integrated into India. Then came the time of the great empires. The Mauryans ruled over almost all the Indian subcontinent for more than a century. They were followed by a string of dynasties, which succeeded one another and coexisted more or less peacefully up to the Muslim Conquest in the eighth century.

Throughout this era, mathematics thrived in India, though sadly we know little about it. The reason for our ignorance is simple: from the start of the Vedic period onwards, its scholars developed an ideal of oral transmission of knowledge, which in principle forbade it to be written down. Knowledge had to be learnt by word of mouth, passed down from generation to generation, from master to pupil. The texts were learnt by heart, in the form of poems or accompanied by clever mnemonics, then recited and repeated as many times as necessary until they were perfectly mastered. Naturally, here and there, one can find a few exceptions to the rule, written fragments that have come down to us, but the harvest is quite meagre.

Nevertheless, they did do maths. How else can we explain the richness of the concepts revealed to the rest of the world when, in around the fifth century AD, the Indians finally decided to put into writing the knowledge they had accumulated orally over centuries? From that point on, India entered a scientific golden age, which soon spread throughout the world.

The Indian scholars set about writing long treatises covering their ancestral knowledge, which they complemented with their own discoveries. The most famous of them included Aryabhata, who was interested in astronomy and calculated very good approximations for the number π; Varahamihira, who produced new advances in trigonometry; and also Bhāskara I, who was the first to write the zero in the form of a circle and to employ the decimal system that we still use today in a scientific way. Indeed, our ten figures, 0, 1, 2, 3, 4, 5, 6, 7, 8 and 9, which we usually refer to as Arabic figures, are in reality Indian.

However, if among all the Indian scholars of that period we had to remember just one, there is no doubt that history would choose Brahmagupta. He lived in the sixth century and was head of the observatory in Ujjain. At that time, the town of Ujjain, situated on the right bank of the Shipra in the centre of modern India, was one of the greatest scientific centres on the subcontinent. Its astronomical observatory had made its reputation and the town was already known to Claudius Ptolemy in the grand days of Alexandria.

In 628, Brahmagupta published his major work, the *Brāhmasphuṭasiddhānta*. This text contained the first complete description of the zero and of the negative numbers, together with their arithmetic properties.

In modern times, the zero and the negative numbers have grown so common in our daily life – to measure the temperature, the height above sea level, or our bank balance – that we almost overlook just how brilliant these ideas are. Their invention was a rare exercise in mental acrobatics, which the Indian scholars were the first to execute to perfection. It is an intellectual delight to achieve an understanding of this process, and all that is subtle and powerful about it, and here we shall dwell for a while on the subject, so as to look more deeply into the chaos that overtook mathematics in centuries to follow.

One of the questions I am asked most frequently when I publicly mention my taste for mathematics is how it originated. How did you come to have such a bizarre interest? Was your passion inspired by a particular teacher? Did you already like maths when you were a child? What triggered such a vocation never ceases to arouse the curiosity of people once immune to the world of numbers.

To be honest, I have to say that I just don't know. As far back as I can remember, I have always liked mathematics, without being able to identify a particular event in my life that might have led me down this path. However, when I delve deeper into my memory, I recall how when new ideas flashed into my head I could be plunged into a state of intellectual jubilation. This happened in particular when I discovered a surprising multiplication property.

I must have been nine or ten years old when one day I was messing around at random on my school calculator and came across a strange result: $10 \times 0.5 = 5$. Multiply the number 10 by 0.5, and you get 5; that's what my calculator, in which I had placed a blind yet also unreasoning trust, was claiming to tell me. How can it be that by multiplying a number, you can obtain another one that's smaller than the first? Isn't a multiplication meant to increase the quantity that you apply it to? Wouldn't that go against the very meaning of the verb 'to multiply'? Wouldn't my trusty calculator have done better to display me a number greater than 10?

It took me some time, several weeks, to think it through steadily, before I managed to get my ideas straight. Things finally clicked the day I thought of representing multiplication geometrically, following in the footsteps of the thinkers of antiquity without knowing it. Take a rectangle of length 10 units and width 0.5. Its area is precisely that of 5 small squares of side 1.

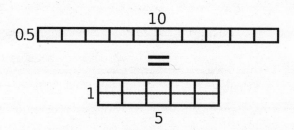

In other words, multiplying by units of 0.5 (as a fraction, ½) comes down to dividing by 2. And the same principle applies to many other numbers. To multiply by 0.25 (¼) is to divide by 4; to multiply by 0.1 is to divide by 10, and so on.

The explanation is convincing, but nonetheless its conclusion still has a disconcerting side: the word 'multiplication' does not mean exactly the same thing when applied in mathematics as it does in everyday language. In day-to-day life, who would claim to have multiplied the area of his garden after selling half of it? Who would say that their fortune has been multiplied after losing 50 per cent of it? On that front, the multiplication of bread becomes a miracle within the reach of everyone: eat half of the loaf and it's done.

When you first come across them, these thoughts tickle your brain. They have something deliciously upsetting about them that resonates in the mind like a particularly well-crafted play on words. At any rate, that was the effect that these curious discoveries had on me as a child. This strangeness appeared all the more clearly to me many years later when, on reading *Science et Méthode*, a text by the mathematician Henri Poincaré published in 1908, I came across the sentence: 'Mathematics is the art of giving the same name to different things.'

To be honest, this phrase can be applied to any language. The word 'fruit' stands for all sorts of different things such as apples, cherries or tomatoes. Each of these words in turn groups together a multitude of different varieties, which in turn provide more subtle categories, and in this way you can move on to an appropriately detailed botanical analysis. However, Poincaré rightly stresses that no language other than that of mathematics goes as far in this process of grouping together. Mathematics allows linkages that no other language permits. For mathematicians, multiplication and division are one and the same operation. Multiplication by one number amounts to division by another. It depends which point of view you adopt.

The inventions of the zero and of the negative numbers derive from the same mindset. In order to create them, you must dare to think outside your own language. Concepts that the language treats in radically different ways have to be grouped together within a single idea. The Indian scholars were the first to embark explicitly on this track.

If I told you that I have already walked on the planet Mars a number of times and have met Brahmagupta in person there on a number of occasions, would you believe me? Probably not. And you would be quite right, because in our language, these sentences mean that I have effectively already walked on Mars and met Brahmagupta. And yet, in mathematics, you only have to imagine that the numbers in the above statement are equal to zero in order to understand that I have not lied. Language uses different structures depending on whether something is or is not the case. Assertion: 'I have walked on Mars'; negation: 'I have not walked on Mars.' Mathematics, on the other hand, erases these differences and groups them together in a single formula. 'I have walked on Mars a certain number of times.' This number can be zero.

While several centuries earlier, the Greeks also had difficulty accepting 1 as a number, imagine what a revolution it was to attribute the noun 'number' to an absence. Before the Indians, a few peoples had latched on to this thought, but none of them had been able to see it through. The Mesopotamians, from the third century BC onwards, had been the first to invent a figure 0. Previously, their numeration system wrote numbers such as 25 and 250 in the same way. Thanks to the figure 0, denoting an empty position, no more confusion was possible. However, the Babylonians never accorded this 0 the status of a number, which could only be written to denote a complete absence of objects.

At the other end of the world, the Mayans had also invented a zero. They even invented two: the first, like that of the Babylonians, was simply used as a figure to mark an empty place in their positional system to base twenty. The second, on the other hand, can certainly be considered to be a number, but was only used in the context of their calendar. Each month of the Mayan calendar had twenty days, numbered from 0 to 19. This zero was used on its own, however; its usage was not very mathematical. The Mayans never used it to perform arithmetical operations.

In short, Brahmagupta was certainly the first to give a complete description of the zero as a number, together with a description of its properties: when any number is taken away from itself, you get zero; when zero is added to or subtracted from a number, that number stays unchanged. These arithmetic properties seem self-evident to us, but the fact that they were so clearly stated by Brahmagupta shows us that the zero was definitively incorporated as a number with a status like all the others.

The zero opened the way to the negative numbers, but it took even longer for mathematicians to adopt these definitively.

Chinese scholars were the first to describe quantities that could be likened to negative numbers. In his commentaries on the *Nine Chapters*, Liu Hui described a system of coloured wands that could be used to represent positive or negative quantities. A red wand denoted a positive number, a black wand a negative one. Liu Hui explained in detail there how these two kinds of numbers interacted with each other, and in particular, how they were added or subtracted.

That description was already very full, but he still had a step to take: that of considering the positive and the negative numbers, not as two distinct groups that can interact, but rather as one and the same entity. Certainly, positive and negative numbers do not always have the same properties when it comes to performing calculations, but above all they have numerous points in common, which allow us to group them together. The situation can be compared to that of even numbers and odd numbers, which form two distinct clans with different arithmetic properties, but which nevertheless belong to the same large family of numbers.

As in the case of the zero, this reunification was first performed by the Indian scholars. And it was again Brahmagupta who provided a complete study in the *Brāhmasphuṭasiddhānta*. In the footsteps of Liu Hui, he established a complete list of the rules that apply to operations with these new numbers. He taught us, among other things, that the sum of two negative numbers is negative, for example $-3 + -5 = -8$; that the product of a positive number and a negative number is negative, $-3 \times +8 = -24$, and again that the product of two negative numbers is positive, $-3 \times -8 = +24$. This last point may appear counterintuitive and has proved one of the most difficult for others to accept. Even today, it is a well-known trap, which schoolchildren worldwide are wary of.

WHY DOES MINUS TIMES MINUS EQUAL PLUS?

In the centuries that followed their announcement by Brahmagupta, the rules for multiplication of signs, and in particular the 'minus × minus = plus', never ceased to arouse distrust and questioning.

This questioning went far beyond the world of mathematicians, and triggered a great deal of incomprehension as soon as the rules began to be taught in schools. In the nineteenth century, the French writer Stendhal himself expressed his incomprehension in his autobiographical novel *Vie de Henri Brulard*. The author of *Le Rouge et le Noir* and *La Chartreuse de Parme* wrote as follows:

MY ENTHUSIASM FOR MATHEMATICS WAS BASED PRINCIPALLY PERHAPS ON MY HORROR OF HYPOCRISY.

WHAT THEN WHEN I REALIZED THAT NO ONE COULD EXPLAIN TO ME HOW IT IS THAT A MINUS TIMES A MINUS EQUALS A PLUS (− × − = +)? (THIS IS ONE OF THE FUNDAMENTAL BASES OF THE SCIENCE KNOWN AS ALGEBRA.) . . .

AT THE AGE OF FOURTEEN, IN 1797, I IMAGINED THAT HIGHER MATHEMATICS, WHICH I HAVE NEVER KNOWN, CONTAINED EVERY OR ALMOST EVERY ASPECT OF OBJECTS, SO THAT BY GOING ON I WOULD COME TO KNOW CERTAIN, INDUBITABLE THINGS, WHICH I COULD PROVE TO MYSELF WHENEVER I WANTED, ABOUT EVERYTHING . . .

I WAS THEREBY REDUCED TO WHAT I STILL TELL MYSELF TODAY: THAT 'MINUS TIMES MINUS EQUALS PLUS' MUST BE TRUE, SINCE SELF-EVIDENTLY, BY CONTINUALLY EMPLOYING THIS RULE IN A CALCULATION, YOU END UP WITH RESULTS THAT ARE TRUE AND INDUBITABLE.

The rule for the multiplication of signs is certainly quite strange at first sight, but it does make sense if you think back to the system of wands devised by the Chinese scholars. For example, let us use this system to represent financial gains or losses. Let us suppose that a black wand represents €5 while a grey wand represents a debt of €5, in other words €−5. Thus, if you have ten black wands and five grey wands, you are €25 in credit.

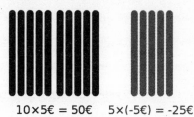

10×5€ = 50€   5×(-5€) = -25€

Let us now study the various scenarios that can arise when your holdings vary. Imagine that someone gives you four more black wands, so that your funds then increase by €20. In other words: 4 × 5 = 20. The product of two positive numbers is indeed positive, and all is well up to this point.

If someone now gives you four grey wands, in other words four debts, your funds decrease by €20. In other words: 4 × −5 = −20. A positive number multiplied by a negative number gives a negative number. And in the same way, if someone takes four black wands away from you, you also lose €20. This amounts to saying that −4 × 5 = −20. These last two situations show clearly that giving someone debts has the same effect as taking money from them. Adding something negative amounts to subtracting something positive.

Now we come to the crucial point: What happens to your funds if someone takes four grey wands away from you. In other words, what happens if someone takes away your debts? The answer is clear: your funds increase, you gain money. And this amounts to saying that −4 × −5 = +20. Taking away something negative amounts to adding something positive – minus times minus equals plus.

The arrival of negative numbers also overturned the meaning of addition and subtraction. The problem is exactly the same as that of multiplication by 0.5, which is division by 2. Since adding a negative number amounts to subtracting a positive number, these two operations lose the meaning they have in everyday language. Addition usually stands for augmentation. However, if I add the number −3, this amounts to removing 3. For example, 20 + −3 = 17. Likewise, if I subtract −3, that amounts to adding 3: 20 − −3 = 23. Once again, we are in the process of giving the same name to different things. Thanks to negative numbers, addition and subtraction become the two faces of one and the same operation.

This confusion of words and the occurrence of what looked like paradoxes, such as the 'minus × minus = plus', put a considerable brake on the adoption of negative numbers. For a long time after Brahmagupta, numerous scholars continued to turn their noses up at these frightfully practical numbers that were also difficult to grasp. Some people called them the 'absurd numbers', and only resigned themselves to using them in their intermediate calculations provided they did not appear in the final result. Their legitimacy was only fully accepted and their use definitively adopted in the nineteenth or even the twentieth century.

In 711, two thousand horsemen and camel drivers from the West turned up in the Indus valley. These troops belonged to Muhammad bin-Qasim al-Sakafi, a young Arab general, who was barely twenty years old. His soldiers, who were better equipped and prepared, routed the 50,000-strong army of the Raja Dahir and seized the region of Sind and the Indus Delta. The event was tragic for the local populations: thousands of soldiers were beheaded and the region was thoroughly pillaged.

The arrival of the new Arab–Muslim Empire at the gates of the Indus resulted in the spread of Indian mathematics. The Arab scholars very quickly incorporated the Indian discoveries into their works, giving them a worldwide resonance that still finds echoes in the mathematics of the twenty-first century.

8

THE POWER OF TRIANGLES

We return to Mesopotamia, where it all began. The year is 762. Although Babylon is already just a field of ruins, enormous construction works are beginning some hundred kilometres further north. It is there, on the right bank of the Tigris, that the Abbasid Caliph Al-Mansur has decided to build his new capital.

At the time, the Arab–Muslim Empire has just undergone a century of rapid expansion. One hundred and thirty years earlier, in 632, when the 34-year-old Brahmagupta had just finished writing the *Brāhmasphuṭasiddhānta*, the Prophet Mohammed was dying in Medina. After his death, the caliphs who succeeded him made a series of conquests and spread Islam from the south of Spain to the banks of the Indus, passing through North Africa, Persia and Mesopotamia.

Al-Mansur ruled over a caliphate with an area greater than 10 million square kilometres. Transposed to today, this territory would be the second-largest country in the world, second to Russia, but bigger than Canada, the United States or China. Al-Mansur was an enlight-ened caliph. To build his capital, he brought in the best architects, artisans and artists of the Arab world. He entrusted the choice of the location and the start date for the building works to his geographers and astrologers.

It took four years and more than one hundred thousand workmen to erect the town of his dreams. A special feature of the town was that it was perfectly round. Its double wall of circular ramparts, with a perimeter of 8 kilometres, was fortified with 112 towers and had four gates, which were oriented on the diagonals at the four cardinal points. The town centre included the barracks, the mosque and the caliph's palace, whose green dome rising to a height of almost 50 metres was visible from all round up to 20 kilometres away.

When it was founded, the town was named Madīnat as-Salām, the town of peace. It also came to be called Madīnat al-Anwār, the town of lights, or again Āsimat ad-Dunyā, the capital of the world. However, Al-Mansur's town entered the history books with another name: Baghdad.

The population of Baghdad rapidly reached several hundred thou-sand. The town was located at the crossroads of major commercial routes, and the streets buzzed with traders from the four corners of the Earth. The stalls were covered with silk, gold and ivory, the air was heady with perfumes and spices, and the town rang with stories from afar. This was the period of the *Thousand and One Nights* and of legends, of sultans, viziers and princesses, and also of magic carpets, of genies and magic lamps.

Al-Mansur and the caliphs who succeeded him wanted to make Baghdad a leading town of culture and science. Thus, in order to attract the greatest scholars, they employed a lure that had already proved itself one thousand years earlier in Alexandria: a library.

At the end of the eighth century, Caliph Harun Al-Rashid began to put together a collection of books with the intention of preserving and bringing to life the knowledge accumulated by the Greeks, Mesopotamians, Egyptians and Indians.

Numerous works were copied and translated into Arabic. The Greek works still in circulation in intellectual circles were the first to be incorporated by the scholars of Baghdad. Within a few years several Arabic editions of Euclid's *Elements* emerged. Several treatises by Archimedes, including that on the measurement of the circle, Ptolemy's *Almagest* and Diophantus' *Arithmetica* were also translated.

At the beginning of the ninth century, the mathematician Muhammad al-Khwārizmī published a major work, *On Calculation with Hindu Numerals*, in which he presented the decimal number system originating in India. Thanks to him, the ten figures, including zero, spread throughout the Arab world, and from there established themselves definitively throughout the world. In Arabic, the zero is called *zifr*, which means 'empty'. With the transfer to Europe, this word took on dual forms: it passed into Italian as *zefiro*, which gave our 'zero'; on the other hand, it became *cifra* in Latin, which gave our word 'cipher'. Unaware of the Indian roots of these ten symbols, the Europeans called them Arabic figures.

In 809, Harun Al-Rashid died and his son Al-Amin replaced him. Al-Amin did not rule for long, being ousted in 813 by his own brother, Al-Ma'mun.

Legend has it that one night Al-Ma'mun was visited in a dream by Aristotle. This encounter made a deep impression on the young caliph, who decided to give a new impetus to scientific research and to welcome even more scholars to his town. Thus, in 832 the library of Baghdad spawned an institution destined to promote the conservation and development of scientific knowledge. The establishment was called Bayt al-Hikma, the House of Wisdom, and the way it worked was strangely reminiscent of that of the Mouseion in Alexandria.

The caliph was heavily involved in its development. He intervened directly with foreign powers, such as the Byzantine Empire, to have rare works sent to Baghdad, which could then be copied and translated there. He commissioned scholarly books intended for distribution throughout the caliphate. He even took part personally in the scientific or philosophical discussions, which were held at least once a week within Bayt al-Hikma.

Over the centuries, Baghdad's House of Wisdom set a popular trend throughout the Arab world. In turn, numerous other towns built themselves libraries and scholarly institutions. Among the most influential and active were Cordoba in Andalusia, founded in the tenth century; Cairo in Egypt, founded in the eleventh century; and Fez in what is now Morocco, founded in the fourteenth century.

It has to be said that this scientific decentralization was broadly enabled by the arrival of an invention – paper – that originated in China and was rediscovered, almost by chance in 751 during the battle of Talas in what is now Kazakhstan. Paper made it easier to copy and transport books. From then on, there was no longer any need to go to Baghdad in order to keep up with the latest discoveries in mathematics, astronomy or geography. Great scientists were able to work and innovate in the four corners of the Arab–Muslim Empire.

THE TILINGS OF THE ALHAMBRA

While the great minds were writing the history of mathematics in Bayt al-Hikma, a different history continued to be made in the streets of Baghdad and the Arab towns. Islam in principle forbids the depiction of human beings or animals in mosques or other religious sites. Thus, to mitigate the effects of this ban, Muslim artists exhibited a breathtaking creativity in the development of decorative geometric patterns.

Do you recall the first sedentary artisans of Mesopotamia, who designed patterns to decorate their pots? Without knowing it, they had found the seven possible categories of friezes. Now, while a frieze is a figure repeated in one direction, you can also imagine figures being repeated in two directions to cover entire surfaces. These are called tilings. The streets of Baghdad and Muslim towns were gradually clothed with a flamboyant geometry that became a trademark of Islamic art.

Some tilings are very simple.

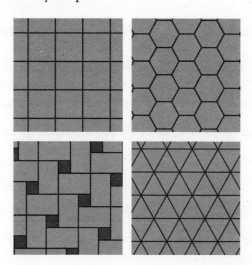

Others are more complicated.

Later, mathematicians were able to prove that there exist precisely seventeen categories of geometric tilings, which are classified according to the geometric transformations that leave them invariant. Each of these categories can then give rise to infinitely many different variants. Without knowing this theorem, the Arab artists discovered the seventeen categories and developed them in a masterful way in their architecture, and also in the ornamentation of *objets d'art* or everyday objects.

In Granada, Andalusia, the Alhambra Palace is one of the most striking monuments attesting to the presence of Islamic art in Spain in the Middle Ages. More than 2 million tourists visit it each year. What few of them know is that the Palace enjoys a very special reputation among mathematicians. The Alhambra is in fact well known for its incorporation of each of the seventeen possible categories of tilings, which are scattered (and sometimes well hidden) throughout its rooms and its gardens.

So, if you should happen to be in Granada one day, you know what to do.

Let us linger in Baghdad and open the doors of the Bayt al-Hikma. What new mathematics are these Arab mathematicians concocting for us? What is the subject of these newly written books that are lining up on the library bookshelves?

One of the disciplines that witnessed the greatest development during this period was trigonometry, that is to say the study of the measurements of trigons, otherwise known as triangles. At first sight, that may seem disappointing: the peoples of antiquity had already studied triangles, as Pythagoras' theorem demonstrates. However, the Arabs extended their research to a point of making a discipline with a remarkable accuracy out of them, whose results still find many applications today.

Contrary to what one might think, triangles are not always all that easy to understand, and many points remained to be clarified at the end of antiquity. To have a good knowledge of a triangle, we essentially need six pieces of information about it: the lengths of its three sides and the sizes of its three angles.

But look: in order to use trigonometry on the ground, it is often much simpler to measure the angle between two directions than the distance between two points. Astronomy is the most striking example of this. To determine the distance separating the stars observed in the night sky is a very difficult question, which took several more centuries to answer. On the other hand, to measure the angle that these stars make between themselves or above the horizon proved easier. A simple octant, an ancestor of the sextant, was all that was needed. Similarly, a geographer who wished to create a map of a region could easily measure the angles of a triangle formed by three mountains. He only needed an alidade, which was just a protractor with a sighting system, for that. And for orienting the map in space, a simple compass enabled him to measure the angle between north and a given direction. Measurement of the distance between the three mountains, on the other hand, called for mounting a much more serious expedition and substantially more complicated calculations. Alexander and his Bematists would not contradict us on this point.

The aim of the game is then as follows: how should you proceed in order to know all the information about a triangle by measuring the fewest distances possible? In asking this question, the trigonometers came up against a problem similar to that which Archimedes had faced when considering the circle a millennium earlier. First, if you know all the angles of a triangle, but none of its sides, you can deduce its shape, but not its size. By way of proof, the following triangles all have the same angles, but the lengths of their sides are different.

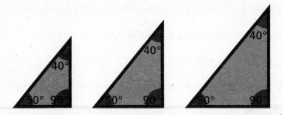

However, they all have the same proportions. If, for example, you ask by how much you have to multiply the length of the longest side by to obtain the shortest one, the result turns out to be the same for each of the three triangles: 0.64! This bears some similarity to the way in which the circumference of a circle is always obtained by multiplying its diameter by π however big the circle is.

Actually, we should have said 'almost 0.64'. This number is just an approximation. As for π, this proportion cannot be calculated precisely and we have to be content with approximate values. A little bit more precision would give us 0.642 or even 0.64278, but this is still not perfect. The decimal expansion for this number has infinitely many figures after the decimal point. The same is true for the other ratios that you can calculate in these triangles. Thus, you obtain the middle side from the longest side by multiplying by approximately 0.766, and the middle side from the small side by multiplying by approximately 1.192.

Since it is impossible to attribute exact values to these three ratios, mathematicians gave them names in order to be able to study them more readily. Several terms were employed according to the places and the periods, but today we refer to them respectively as 'cosine', 'sine' and 'tangent'. A number of variants were also invented and used before being forgotten. One example is the *seked*, which the Egyptians used to measure the slope of their pyramids. Another example is the chord, which was introduced by the Greeks, and corresponded to a ratio in an isosceles triangle.

The trigonometric ratios raised another problem. Their values vary from one triangle to another. Thus, the ratios 0.642, 0.766 and 1.192 are valid only for triangles with the angles 40°, 50° and 90°. If, on the other hand, you look at a right-angled triangle with angles 20°, 70°

and 90°, then its cosine, sine and tangent will be approximately 0.342, 0.940 and 2.747.

In short, the job of mathematicians of trigonometry was much more extensive than was envisaged. It was not simply a matter of finding a number, nor even three; there was a need to calculate tables of numbers that vary as a function of all the possible angles you might need to calculate.

Below we give a trigonometric table for all right-angled triangles for which one of the angles ranges in steps of 10° from 10° to 80°. You will note that a single angle is given for each triangle. In fact, it is not necessary to indicate the other two, which can be easily determined: first, the right angle always measures 90°; and second, a theorem asserts that the sum of the three angles of a triangle is always equal to 180°, and this allows us to deduce the third angle. To tell the truth, it is not even necessary to draw the triangles: a knowledge of the angle alone is sufficient to reconstitute them. This is why the first column of trigonometric tables in general only indicates the angle. Thus, the cosine of the angle 10° is equal to 0.9848 and the tangent of 50° is equal to 1.1918.

TRIANGLE	COSINE	SINE	TANGENT
10°	0.9848	0.1736	0.1763
20°	0.9397	0.3420	0.3640
30°	0.8660	0.5	0.5774
40°	0.7660	0.6428	0.8391

TRIANGLE	COSINE	SINE	TANGENT
50°	0.6428	0.7660	1.1918
60°	0.5	0.8660	1.7321
70°	0.3420	0.9397	2.7475
80°	0.1736	0.9848	5.6713

Of course, a trigonometric table is never complete. It is always possible to refine it, either by finding better approximations for the ratios it contains, or by refining the range of triangles represented. In the table, the triangles have angles varying from 10° in steps of 10°, but it would preferable to have a precision of up to a degree, or a tenth of a degree. In short, the calculation of ever finer trigonometric tables is a never-ending task, and one that generations of mathematicians have tackled, each in their turn. Only the arrival of electronic calculators in the twentieth century eventually relieved them of their burden.

The Greeks were undoubtedly the first to compile trigonometric tables. The oldest ones that have come down to us are to be found in Ptolemy's *Almagest* and are said to derive from Hipparchus of Nicaea, a mathematician of the second century BC. At the end of the sixth century, the Indian scholar Aryabhata also published his tables of trigonometry. In the Middle Ages, the most famous tables were compiled by the Persians, Omar Khayyam in the eleventh-twelfth century and Al-Kashi in the fourteenth century.

The scholars of the Arab world played a primordial role, not only through their contribution to the compilation of more precise tables, but also and above all through what they did with them. They took the art of juggling with this data and using it as effectively as possible to its peak.

Thus, in 1427 Al-Kashi published a work entitled *Miftah al-hisab*, or *Key of Arithmetic*, in which he stated a result that generalized Pythagoras' theorem. By skilful use of cosines, Al-Kashi managed to produce a theorem that applied to absolutely all triangles, and no longer solely to right-angled triangles. Al-Kashi's theorem (now known as the cosine rule) was based on a correction to Pythagoras' theorem: when the triangle is not right-angled, the sum of the squares of the first two sides is not equal to the square of the third side; however, this equality becomes true if you add a correction term that is calculated directly from the cosine of the angle between the first two sides.

When Al-Kashi published this result, he was already no longer a nobody in the mathematical world. He had become known three years earlier for his calculation of an approximation of the number π up to the sixteenth decimal digit – a record for the period. But records are made to be broken.* Theorems, on the other hand, remain. To this

---

* The Dutch mathematician Ludolph Van Ceulen calculated 35 decimal digits one hundred and seventy years later.

day, Al-Kashi's theorem is still one of the most commonly used trigonometric results.

The scene is the left bank of the Seine in Paris. It is June, and I have
been transformed into a rather special tour guide. Today, with a group
of some twenty people, we are walking the streets of the Latin Quarter
on the trail of mathematics and its history. Our next scheduled stop
is in the Jardin des Grands Explorateurs (Garden of the Great
Explorers). To the north, we can see the symmetric paths of the Jardin
du Luxembourg fleeing in massed ranks towards the Senate building.
To the south, the rounded silhouette of the dome of the Paris
Observatory rises up over the roofs of the capital.

Like tightrope walkers, we tread along the exact line of the Paris
Meridian, following the axis of the garden. One step out to the left
and we are in the eastern hemisphere of the world. Two steps to the
right and we swing into the western hemisphere. Five hundred metres
further on, the Meridian runs through the centre of the Observatory,
passes through the middle of the 14th arrondissement and then leaves
Paris via the Parc Montsouris. It follows its course through the French
countryside, cuts into part of Spain and sets off across the African
continent and the Antarctic Ocean to end at the South Pole. Behind
us, it climbs the streets of Montmartre, and brushes past the British
Isles and Norway before reaching the North Pole.

Establishing the exact course of the Meridian was no easy matter.
It required precise surveys over vast areas. For example, how can you
measure the distance between two points located on the opposite sides
of a mountain you are unable to cross? To answer this question, the

scholars of the beginning of the eighteenth century embedded the Meridian in a succession of virtual triangles running from the North to the South of France. The anchor points for the triangulation were chosen to be points at an altitude, such as hills, mountains or bell towers, from which it was possible to take a bearing on the other points to measure the angles between them. Once the survey data had been collected on the ground, it merely remained to make extensive use of the trigonometric procedures perfected by the Arabs to determine the exact position of each of the triangulation points, and through these that of the Meridian.

The Cassinis were among the first to devote themselves to this task. The Cassini family was a veritable dynasty of scientists, to the point where they were usually given a number, as with kings. Giovanni Domenico, known as Cassini I, a recent émigré from Italy, was the first director of the Paris Observatory from its foundation in 1671. His son Jacques, or Cassini II, succeeded him on his death in 1712. It was they who established the first triangulation of the Meridian, which was completed in 1718. After them, Cassini III (forenames César-François and son of II) turned their triangulation of the Meridian into the spinal column of the first complete triangulation of the French territory. This led to the publication in 1744 of the very first map of France established by a rigorous scientific procedure. His son Cassini IV, forenames Jean-Dominique, continued his work by refining the triangulation further region by region.

As we walk along the Meridian, we are walking indirectly in the footsteps of the Arab scholars who established the theoretical bases for these triangulations. Each triangle on the map required the use of cosines, sines or tangents. In their shape, each of them carries the

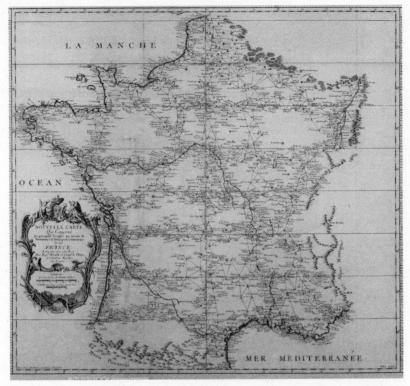

*Map of France of 1744 showing the Paris Meridian
and the main Cassini triangles*

legacy of Al-Kashi and the first trigonometers of Baghdad. All these calculations were done by hand, and took countless hours of work on the part of the scholars of the Observatory with their trigonometric tables.

Triangulation continued to be used until the end of the twentieth century and the advent of satellites. The most precise grids then included up to 80,000 points. The stones that marked these points can still be seen, scattered almost everywhere throughout the French territory. In Paris, you can still see the two triangulation pillars that determined the axis of the Meridian: one is in the south in the Parc

Montsouris, the other is in the north in Montmartre. In 1994, 135 medallions named after the astronomer François Arago were placed along the path of the Meridian in the capital. One of these is actually inside the Musée du Louvre. The next time you are out for a stroll in the streets of Paris, keep your eyes open, you might well come across some.

When the metric system was introduced at the time of the French Revolution, the length of the metre was related, in a gesture of universality, to that of the Meridian. One metre was precisely defined as one ten-millionth part of one quarter of the Meridian. In 1796, sixteen standard metres, engraved in marble, were installed in all corners of Paris so that citizens could go and refer to them. Today, two of these can still be seen, one on Rue Vaugirard facing the Jardin du Luxembourg, the other in the Place Vendôme at the entrance to the Ministry of Justice.

The Paris Meridian acted as a reference until the International Meridian Conference in Washington, DC in 1884. It was then replaced by the Greenwich Meridian, which passes through the Royal Observatory in London. In exchange for the Meridian, the British committed themselves to adopting the metric system. Matter on hold.

With the arrival of information technology and satellites, trigonometric tables and triangulations on the ground lost their usefulness. But for all that, trigonometry has not disappeared. It has come to reside in the heart of processors. The triangles are hidden, but they are still there.

Nearby cars are heading up the Avenue de l'Observatoire. Many of them are now fitted with a Global Positioning System (GPS). At every instant, their paths are determined by their positioning relative to four satellites that track them from space. The solution of the resulting equations again calls upon trigonometry. Do the drivers of these cars know that the voice that calmly orders them to turn left has just that very instant used a few sines or cosines?

Have you ever heard one of the investigators from a detective TV programme say that the suspect's phone was just located by triangulation? This type of positioning involves determining the position of a mobile as a function of its distance from the three nearest relay antennas. This problem in geometry is readily solved using a few formulae from trigonometry, which our computers now work out at lightning speed.

And, not content to measure real things, trigonometry also intrudes on the creation of virtual worlds. 3D animated films and video games make extensive use of it. Beneath the texture with which the graphic designers cover them, 3D shapes consist of geometric meshes curiously reminiscent of the Cassini triangulations. It is the deformation of these meshes that animates the objects and the people. The calculation of even the smallest synthesized image, such as that of the Utah teapot that was one of the first objects to be modelled on a computer in 1975, requires the application of a large number of trigonometric formulae.

9

INTO THE UNKNOWN

Back to Baghdad. Of all the scholars who frequented the Bayt al-Hikma, one of them in particular left his mark on his era: Muhammad ibn al-Khwārizmī.

Al-Khwārizmī was a Persian mathematician born in the 780s. His family originated in the Khwarezm region that extends over present-day lands in Iran, Uzbekistan and Turkmenistan, and no one really knows whether al-Khwārizmī was born there or whether his parents emigrated to Baghdad before he was born. In any case, the young scholar was to be found in that round town at the start of the ninth century. He was one of the first scientists to enter Bayt al-Hikma and make a first-class reputation there.

On the streets of Baghdad, al-Khwārizmī was primarily known as an astronomer. He wrote several theoretical treatises on the knowledge of the Greeks and the Indians, as well as practical works on the use of a sundial or the construction of an astrolabe. He also put his knowledge to good use to compile geographical tables listing the latitudes and longitudes of the most remarkable places in the world. However, his reference meridian, inspired by Ptolemy, remained approximate: it was defined as passing through the Fortunate Islands, whose more or less mythological location is thought to have been in the far west of the world, and that may correspond to the present-day Canary Islands.

In mathematics, al-Khwārizmī was the author of the famous book *On Calculation with Hindu Numerals*, which revealed the decimal positional system to the world. This essential work alone would have guaranteed him a place in the hall of fame of mathematics. However, it is another book with revolutionary content that definitively earned him a place among the greatest mathematicians in history, alongside Archimedes and Brahmagupta.

It was al-Mamun in person who commissioned this book from him. The caliph wished to offer his subjects a manual of mathematics for use by anyone and everyone to solve questions that cropped up in daily life. Al-Khwārizmī was entrusted to produce it, and began to compile a list of classical problems together with the methods used to solve them. This included, among other things, questions about the measurement of land, commercial transactions, and also the distribution of a legacy among the different members of a family.

While all these problems were very interesting, there was nothing innovative about them, and had al-Khwārizmī stuck to the caliph's request, his book would undoubtedly never have been passed down to posterity. But the Persian scholar did not stop there, and decided to add a purely theoretical first part in the introduction to his work. There he presented the various methods of solutions, which were put into practice in concrete problems in a structured and abstract way.

Having completed the work, al-Khwārizmī gave it the title *Kitāb al-mukhtaṣar fī ḥistāb al-jabr wa-l-muqtābala*, or *The Compendious Book on Calculation by Completion and Balancing*. When it was translated into Latin, much later on, the last words of the Arabic title were transcribed phonetically and the book was called *Liber algebræ* and *Almucabola*. Gradually the term *Almucabola* was discarded, leaving the single word that from then on denoted the discipline initiated by al-Khwārizmī: *al-jabr, algebræ . . . algebra*.

It was not so much its mathematical content, but rather al-Khwārizmī's formulation of his methods that was revolutionary. He detailed his problem-solving procedures independently from the problems themselves. For a better understanding of this approach, let us look at the following three questions:

1. **A RECTANGULAR FIELD IS 5 UNITS WIDE AND HAS AN AREA OF 30. WHAT DOES ITS LENGTH MEASURE?**

2. **A MAN OF AGE 30 YEARS IS 5 TIMES THE AGE OF HIS SON. WHAT IS THE AGE OF HIS SON?**

3. **A TRADER HAS BOUGHT 30 KILOGRAMMES OF FABRIC ON 5 IDENTICAL ROLLS. HOW MUCH DOES EACH ROLL WEIGH?**

In all three cases, the answer is 6. And in solving these problems you have the feeling that, although they deal with radically different subjects, the mathematics hidden behind them is the same. In these three cases, the result is found by division: $30 \div 5 = 6$. Al-Khwārizmī's initial approach involved stripping these questions of their context in order to extract a purely mathematical problem from them:

FIND A NUMBER THAT, MULTIPLIED BY 5, GIVES 30.

In this formulation, we do not know what the numbers 5 and 30 represent. They may be geometric dimensions, ages, rolls of fabric or anything else – no matter, that doesn't change anything about the way that we look for the answer. The aim of algebra is thus to propose methods that make it possible to solve purely mathematical riddles of this type. In Europe, several centuries later, these riddles became known as equations.

Al-Khwārizmī went even further in his study of equations. He asserted that the method does not even depend on the numerical data of the problem. Let us consider the following three equations:

1. FIND A NUMBER THAT MULTIPLIED BY 5 GIVES 30;

2. FIND A NUMBER THAT MULTIPLIED BY 2 GIVES 16;

3. FIND A NUMBER THAT MULTIPLIED BY 3 GIVES 60.

Each of these equations also combines, in its formulation, a multitude of different concrete problems. But, once again we sense that they will all be solved using the same method. In these three cases, the solutions are found by dividing the second number by the first: for the first, $30 \div 5 = 6$, for the second, $16 \div 2 = 8$, and for the third, $60 \div 3 = 20$. The method of solution is thus not only independent from the concrete problem, but also from the numbers that occur in that problem.

It thus becomes possible to formulate these equations in an even more abstract way:

FIND A NUMBER THAT, MULTIPLIED BY A CERTAIN QUANTITY 1,

GIVES A QUANTITY 2.

All problems of this type will be solvable in the same way: you have only to divide quantity 2 by quantity 1.

But, of course, this example is still very simple. It only involves a multiplication, and its solution only uses a division. But one can imagine other types of equations in which the unknown undergoes several different operations. Al-Khwārizmī worked mainly on equations in which the unknown may undergo the four basic operations (addition, subtraction, multiplication and division), and also squaring. Here is an example:

FIND A NUMBER WHOSE SQUARE IS EQUAL TO 3 TIMES ITS VALUE, AUGMENTED BY 10.

This time, the solution is 5. The square of 5 is 25 and we have 25 = 3 × 5, + 10, as required. On this occasion, we have been lucky because this solution is a whole number – an integer – and it would have been possible to guess it by having several attempts. But when the solutions are very large, or numbers with a decimal point, it becomes necessary to have a precise method that enables you to find their values in a systematic manner. This is exactly what al-Khwārizmī constructed in the introduction to his book. There, he described step by step the calculations that had to be performed based on the data of the problem, and what these data were. He then also gave proofs to demonstrate that his methods worked.

Al-Khwārizmī's approach thus fitted perfectly with the global dynamics of mathematics, which tends to abstraction and generality. For a long time before then, mathematics had been made independent of the real objects it represented. With al-Khwārizmī, it was the reasoning that was done about these objects that itself became detached from the problems it was intended to solve.

THE CLASSIFICATION OF EQUATIONS

Not all equations are equally easy to solve. There are even some that present-day mathematicians are still finding hard to crack. The difficulty of an equation essentially depends on the operations it involves.

Thus, if the unknown only undergoes additions, subtractions, multiplications and divisions, we speak about equations of degree one. Here are a few examples:

WHICH NUMBER GIVES 10 IF YOU ADD 3 TO IT?

WHICH NUMBER GIVES 15 IF YOU DIVIDE IT BY 2?

WHICH NUMBER GIVES 0 IF YOU MULTIPLY IT BY 2 THEN SUBTRACT 10 FROM IT?

Equations of degree one are the easiest to solve. With a little thought you can find the solutions of these three: 7 since $7 + 3 = 10$, and 30 since $30 \div 2 = 15$, and finally 5 since $5 \times 2 - 10 = 0$.

If you add squaring to these operations, that is to say the operation involving multiplying the unknown by itself, you then move to equations of degree two, and the difficulty increases. It is precisely these equations of degree two that al-Khwārizmī solved in his work. Here are two examples that he dealt with:

THE SQUARE OF A NUMBER PLUS 21 IS EQUAL TO 10 TIMES THAT NUMBER.

THE SQUARE OF A NUMBER WITH 10 TIMES THAT NUMBER ADDED TO GIVES 39.

One feature of equations of degree two is that they can have two solutions. This is the case here: the numbers 3 and 7 (with squares 9 and 49) correspond to the first question, since $3 \times 3 + 21 = 3 \times 10$ and $7 \times 7 + 21 = 7 \times 10$. The second equation also has two solutions: 3 and −13.

In the ninth century, geometry was still the major discipline in mathematics, and al-Khwārizmī's proofs were systematically formulated in geometric terms. According to the interpretation introduced by the scholars of antiquity, the square of a number and the result of multiplying two numbers can be viewed as areas. An equation of degree two can therefore be treated as a problem of plane geometry. Here, for example, are the geometric versions of our last two equations. The question marks are the lengths that correspond to the unknown number.

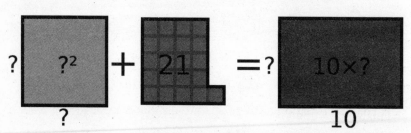

The square of a number plus 21 is equal to 10 times that number.

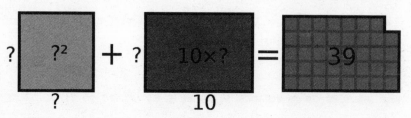

The square of a number with 10 times that number added to it gives 39.

Al-Khwārizmī then solved these problems using an enhancement of the jigsaw-puzzle methods. He carved up the pieces, and added or removed fragments as required to obtain a figure on which the solution was apparent.

Let us look, for example at the second of the above equations. His method begins by cutting up the rectangle equal to ten times the unknown into two rectangles each equal to five times the unknown.

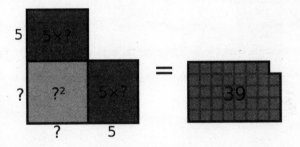

He then pastes the fragments together as follows.

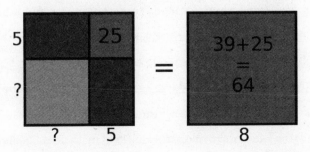

Finally, he adds a piece with an area of 25 to both sides of the equation in order to complete the squares on both sides.

The square on the left then has a side equal to the unknown augmented by 5, while that on the right has a side equal to 8. It follows that the unknown is equal to 3.

Note that the above figure is grossly out of proportion. You could not know before solving it that the unknown was equal to 3, and the lengths represented are incorrect. That does not matter at all, because it is not the numerical values that count here, but the fact that the same cut-and-paste operations work whichever particular numbers appear in these equations. There is a saying that geometry is the art of correct reasoning on incorrect figures – this is a perfect illustration.

However, we note that in this method, the unknown is a length, that is to say a positive number: negative solutions vanish into oblivion. Although our equation has a solution equal to −13, al-Khwārizmī ignored it completely.

After degree two comes degree three. This time, the cube of the unknown may be involved. These equations were too complicated for al-Khwārizmī, and were only solved in the Renaissance. If we interpret them in geometric terms, we then have a problem of volumes in three dimensions.

Then come equations of degree four. From a numerical point of view, the occurrence of these equations is not a concern. However, their geometric representation evades us, for we would have to imagine figures in four dimensions, which is something that cannot be envisaged in our limited three-dimensional world.

This ability of algebra to generate problems that are a priori inaccessible to geometry was to a large extent responsible for the shift of emphasis that took place during the Renaissance and that saw the former rob the latter of the title of the queen of mathematical disciplines.

At the end of the ninth century, the Egyptian mathematician Abu Kamil was one of the main successors to al-Khwārizmī. He generalized al-Khwārizmī's methods, and was interested, in particular, in systems of equations. These systems involve finding several unknown numbers simultaneously from several equations. Here is a classical example.

A camel driver's caravan consists of dromedaries, which have one hump, and camels, which have two. Within the caravan we can count a total of 100 heads and 130 humps. How many animals of each species are there?

Here we are looking for two unknowns, the number of dromedaries and the number of camels, and the information available to us is mixed. The heads and the humps give us two equations, but it is not possible to solve these two equations independently: the problem has to be considered as a whole.

There are several means of tackling this problem. One way is to argue as follows. Since there are 100 heads, there are 100 animals. Now, if there were only dromedaries, there would also be 100 humps and so there would be 30 humps fewer. Thus, there are 30 camels and the 70 others are dromedaries. There is only one solution here, but other more complicated systems may have many more. Indeed, Abu Kamil claimed in one of his works to have solved certain equations for which he found 2,676 different solutions!

In the tenth century, Al Karaji was the first to write that one can conceive of equations of any degree whatsoever, although the scenarios he succeeded in solving were relatively sparse. In the eleventh and twelfth centuries, Omar Khayyam and Sharaf al-Dīn al-Tūsī tackled equations of degree three. They managed to solve certain special cases, and their study produced significant advances without, however, coming up with a systematic method of solution. Several other efforts failed, and some mathematicians began to consider the possibility that these equations might not be solvable.

Ultimately, it was not the Arab scholars who resolved the question. In the thirteenth century, the Golden Age of Islam had already seen its finest years and was in slow decline. There were various reasons for this decline: the domination of the Arab–Muslim Empire did not discourage greed and was regularly attacked, both commercially and militarily.

In 1219, the Mongol hordes of Genghis Khan swept into the Khwarezm region from which al-Khwārizmī originated. In 1258 they were at the gates of Baghdad under the command of Hulagu Khan, grandson of Genghis. Caliph Al-Musta'sim had to capitulate. Baghdad was pillaged and torched and its inhabitants massacred. In the same period, the Reconquista of the territories of southern Spain by the Christian peoples was gathering speed. Cordoba, the capital of the region, fell in 1236. Spain was entirely reconquered in 1492 with the fall of Granada and its Alhambra Palace.

The scientific organization of the Arab world was decentralized enough to resist these defeats for a certain time. First-class research continued to be carried out within it right up to the sixteenth century, but the tide of history was turning, and Europe was preparing to take up the torch of mathematics.

10

IN SEQUENCE

In the medieval period, it has to be said, mathematics did not thrive in Europe. There were, however, a few exceptions. The greatest European mathematician of the Middle Ages was undoubtedly the Italian Leonardo Fibonacci, who was born in Pisa in 1175 and died in the same town in 1250.

How could you become an important mathematician at that time in Europe? The simple answer was, by not staying there. Fibonacci's father was a representative for traders of the Republic of Pisa in the port of Bugia (now known as Bejaia) in what is now Algeria. It was there that Leonardo received his education and discovered the works of the Arab mathematicians, and in particular those of al-Khwārizmī and Abu Kamil. Back in Pisa, in 1202 he published the *Liber abaci*, the *Book of Calculating*, in which he presented an extensive range of mathematics of the period, from Arabic figures to Euclidean geometry, via the results of Diophantine arithmetic and the calculations of numerical sequences. In fact, it was a numerical sequence that ensured his great popularity in the centuries that followed.

A numerical sequence is a regular succession of numbers that can be extended *ad infinitum*. We are already familiar with some of these. The sequence of odd numbers (1, 3, 5, 7, 9 . . .) and that of the squares (1, 4, 9, 16, 25 . . .) are among the simplest examples. In one of the problems in the *Liber abaci*, Fibonacci sought to model the evolution of a population of rabbits mathematically. He considered the following simplified hypotheses:

1. **A PAIR OF RABBITS IS NOT OF REPRODUCTIVE AGE FOR ITS FIRST TWO MONTHS.**

2. **FROM ITS THIRD MONTH ONWARDS, A PAIR GIVES BIRTH TO A NEW PAIR EVERY MONTH.**

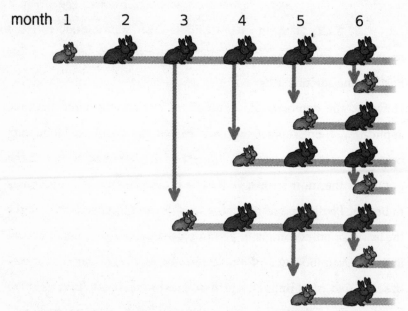

Each row represents the evolution of a pair of rabbits over time.
The unshaded rabbits represent new births.

From these hypotheses, one can predict the descendant tree of a pair of young rabbits.

One can then look at the sequence formed by the number of pairs over time. Looking at the above tree, column by column, we find the values for the first six months: 1, 1, 2, 3, 5, 8 . . .

Fibonacci remarked that each month, the population of rabbits was equal to the sum of the two previous months: $1 + 1 = 2$; $1 + 2 = 3$; $2 + 3 = 5$; $3 + 5 = 8$ . . . and so on. This rule can be explained as follows. Every month the number of pairs that are born, and thus add to the rabbits already present, is equal to the number of pairs of reproductive age in the previous month, in other words to the number of pairs that were already born two months earlier. It is then possible to calculate the terms of the sequence without having to delve into the precise details of the rabbits' genealogy.

1, 1, 2, 3, 5, 8, 13, 21, 34, 55, 89, 144 . . .

For Fibonacci, this problem was primarily a recreational puzzle. However, the demographic sequence of the rabbits found multiple applications, both practical and theoretical, over the following centuries.

One of the most striking examples is undoubtedly its appearance in botany. Phyllotaxy, or phyllotaxis, is the discipline that studies how the leaves or other component elements of a plant are arranged around its axis. If you look at a pine cone you will see that its surface consists of scales that wind round in spirals. More precisely, you can count the number of spirals that turn in a clockwise direction and those that turn in the opposite direction.

8 spirals   13 spirals

As surprising as it may seem, these two numbers are always two consecutive terms of the Fibonacci sequence. When you go for a walk in a forest you may find, for example, pine cones of type 5-8, 8-13 or 13-21, but never 6-9 or 8-11. These Fibonacci spirals appear, more or less clearly, in numerous other plants. While they are clearly visible in pineapples or in the flower heads of sunflowers, they are on the other hand a lot more difficult to detect in the bulging shape of a cauliflower. Yet there they are.

THE GOLDEN RATIO

Among other curiosities, the Fibonacci sequence also revealed a very deep connection with a number that had been known since antiquity: the golden ratio. Its value is approximately equal to 1.618, and the Greeks considered it to be a perfect proportion. Just as for the number π, the golden ratio has an infinite decimal expansion, which is why it was given a name, φ, pronounced 'phi'.

The golden ratio crops up in numerous geometrical variants. A golden rectangle is one whose length is φ times greater than its width. The properties of the golden ratio mean that if you cut out a square based on the rectangle's width, then the small rectangle that remains is still a golden rectangle.

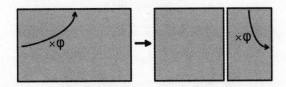

The Greeks used this, particularly in their architecture. The façade of the Parthenon in Athens has proportions very similar to the golden rectangle, and even though it is difficult to find reliable sources about the wishes of the architects, it is highly likely that this is no accident. The first text that has come down to us that defines the golden ratio explicitly is Book VI of Euclid's *Elements*.

It also puts in an appearance in regular pentagons: their diagonals and their sides are in precisely the golden ratio. In other words, the length of one of the five diagonals is equal to the length of a side multiplied by φ.

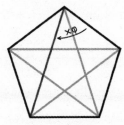

Thus, the golden ratio is to be found in all geometric structures that involve pentagons. This is, for example, the case for the geode or for the footballs that we met earlier. When you seek to calculate its exact value by algebraic methods you come across the following equation of degree two.

The square of the golden ratio is equal to the golden ratio augmented by one.

Al-Khwārizmī's method of completing the square can then be used to obtain the exact formula for it. This gives $\varphi = (1 + \sqrt{5}) \div 2 \approx 1.6180341$.* You can then check that this value does in fact satisfy the equation: $1.618034 \times 1.618034 \approx 2.618034$.

But what does the Fibonacci sequence have to do with this?

If you observe the multiplication of the rabbits for a long enough time, you find that every month, their number is approximately multiplied by φ! Let us look, for example, at the fifth and sixth months. The population rises from 8 to 13 rabbits – it has thus been multiplied by $13 \div 8 = 1.625$. Now certainly, this is not very far from the golden ratio, but it is not exactly that either. If we now look at the transition from the eleventh to the twelfth month, the population is multiplied by $144 \div 89 = 1.61797\ldots$ This is getting closer. And we could carry on. The more time passes, the closer to the golden ratio the multiplicative factor from one month to the next becomes!

Once you notice this, then comes the time for questions. Why? How is it that this anodyne-looking number is present in three distinct domains of mathematics: geometry, algebra and sequences? We might have thought, at the beginning, that we were dealing with three similar but different numbers. But no: the more precisely we measure the diagonal of a pentagon, the more finely we calculate $(1 + \sqrt{5}) \div 2$, and the further we go in the Fibonacci sequence, the more we must bow to the obvious fact that we are dealing with the same number in each case.

To answer this question, mathematicians needed to produce mixed proofs, which created bridges between different branches of

---

* The notation $\sqrt{5}$ in this formula denotes the square root of the number 5, which is to say the positive number whose square is equal to 5. This number is approximately equal to 2.236. The useful symbol $\approx$ means 'is approximately equal to'.

mathematics. This phenomenon, which already existed between geometry and algebra through the figurative representations of numbers in antiquity, now spread to other branches of mathematics. Certain disciplines that had once seemed far removed got into conversation. Beyond their particular interest, numbers such as φ were revealed as formidable go-betweens. In Fibonacci's day, the number π still did not extend its reach beyond geometry, but in the centuries that followed it was that number that became the undisputed champion of all these bridging numbers.

The study of sequences also made it possible to throw new light on the paradoxes of Zeno of Elea, and in particular on that of Achilles and the tortoise. You will recall the race imagined by the Greek scholar: the tortoise starts the race with a lead of one hundred metres over Achilles, but the latter runs twice as fast. In this situation, the paradox seemed to show that despite its slowness the tortoise could never be passed.

This conclusion arose from a breakdown of the race into an infinite number of stages. When Achilles reaches the starting point of the tortoise, the latter will have advanced by 50 metres. In the time that Achilles runs these 50 metres, the tortoise will be 25 metres further on, and so on. The gaps between the two runners at each of these stages form a sequence in which each term is equal to one half of the previous one.

100 50 25 12.5 6.25 3.125 1.5625 . . .

The sequence is infinite, and that is why one might wrongly deduce from it that Achilles will never catch the tortoise. However, if we add these infinitely many numbers, we find a result that is not at all infinite.

$$100 + 50 + 25 + 12.5 + 6.25 + 3.125 + 1.5625 + \ldots = 200.$$

This is one of the greatest curiosities of sequences: the sum of infinitely many numbers can be finite. The above sum shows us that Achilles will overtake the tortoise at the end of 200 metres.*

These infinite additions also prove very useful for calculating numbers that arise in geometry, such as π or the trigonometric ratios. While these numbers cannot be expressed using the classical elementary operations, it becomes possible to obtain them through sums of sequences. One of the first to explore this possibility was the Indian mathematician Madhava of Sangamagrama, who in around AD 1400 discovered a formula for the number π:

$$\pi = \left(\frac{4}{1}\right) + \left(-\frac{4}{3}\right) + \left(\frac{4}{5}\right) + \left(-\frac{4}{7}\right) + \left(\frac{4}{9}\right) + \left(-\frac{4}{11}\right) + \left(\frac{4}{13}\right) + \cdots$$

The terms of Madhava's sequence are alternately positive and negative, and are obtained by dividing 4 by the successive odd numbers. However, let it not be thought that this sum settles the problem of π definitively. Once the addition has been formulated, we still have to find the result. And while certain sums of sequences, like that of Achilles and the tortoise, can be easily calculated, others are particularly resistant, as is the case for Madhava's sequence.

In short, this infinite sum does not really allow us to write down an exact decimal expansion for π, but it opens new doors for better approximations. Since it is not possible to add infinitely many terms

---

* The calculation of the sum of infinitely many numbers uses the notion of the limit The method involves truncating the sum to look at only a finite number of terms and then adding in more and more to see the limit that these truncated sums approach In the case of Achilles and the tortoise, if you only look at the first seven terms, you find: 100 + 50 + 25 + 12.5 + 6.25 + 3.125 + 1.5625 = 198.4375. If you extend the sum up to the twentieth term, you find approximately 199.9998. It can be shown that as you add more and more terms, the result does indeed converge to 200.

all at once, we can always limit ourselves to just taking a finite number. Thus, keeping only the first five terms, we find 3.34.

$$\left(\frac{4}{1}\right) + \left(-\frac{4}{3}\right) + \left(\frac{4}{5}\right) + \left(-\frac{4}{7}\right) + \left(\frac{4}{9}\right) \approx 3.34.$$

This is not a very good approximation, but no matter, let us go further. Taking the first hundred terms, we arrive at 3.13 and after a million terms we reach 3.141592.

But surely, it is not very practical to add a million terms in order to obtain an approximation to only six decimal places? Madhava's sequence has the disadvantage of converging very slowly. Later, other mathematicians such as the Swiss Leonhard Euler in the eighteenth century or the Indian Srinivas Ramanujan in the twentieth century discovered a host of other sequences whose sum is equal to π, but which converge much faster. These methods gradually replaced the method of Archimedes and made it possible to calculate more and more decimal places.

The trigonometric ratios also have their sequences. Here, for example, is the sum for the cosine of a given angle:

$$\text{cosine} = 1 - \frac{\text{angle}^2}{1 \times 2} + \frac{\text{angle}^4}{1 \times 2 \times 3 \times 4} - \frac{\text{angle}^6}{1 \times 2 \times 3 \times 4 \times 5 \times 6} + \cdots$$

To find the value of the cosine, it suffices to replace 'angle' (see above) by the size of the angle in question*. Similar formulae exist for sines

---

* Note, however, that in order for the formulas to work, the angle should not be measured in degrees, but in radians. With this new unit a full turn is not 360 degrees, but 2π radians. This may seem strange, but the trigonometric formulae and their associated sequences only work correctly with this unit.

and tangents and also for a myriad of other special numbers occurring in different contexts.

Today, sequences continue to have many and varied applications. In the footsteps of Fibonacci, they are still used in population dynamics to study the evolution of animal species over time. However, current models are much more precise, and take account of various other parameters such as mortality, predators, climate or, more generally, the variability of the ecosystems in which the animals live. In the broader picture, sequences are involved in the modelling of any process that evolves step by step over time. Information technology, statistics, economics and meteorology are just some of the fields that call upon them.

11

IMAGINARY WORLDS

At the start of the sixteenth century, the seeds sown by Fibonacci began to bear fruit with the emergence of a new generation of mathematicians. These took up in their own right the algebraic research begun by the Arab scholars. It was they who finally settled the matter of equations of degree three in the culmination of a very weird affair.

This story begins at the beginning of the sixteenth century with a businessman and professor of arithmetic at the University of Bologna by the name of Scipione Del Ferro. Del Ferro was interested in algebra, and he was the very first to discover the formulae for solving equations of degree three. Alas, at that time, the spirit of dissemination of knowledge, which had reigned in the Arab world, had not yet caught on in Europe. The University of Bologna regularly renewed its various professorial positions. In order to stay on top and to keep his place, Del Ferro had every interest in making sure that his competitors did not know his secret. He wrote down his discovery, but he did not publish it. He simply revealed it to a handful of disciples who, as he himself did, kept it confidential.

When Del Ferro died in 1526, the Italian mathematical community therefore still had no idea that equations of degree three had been solved. Many of its members even continued to think that they were simply not solvable. However, a disciple of Del Ferro by the name of Antonio Maria Del Fiore, in whom his master had confided, could not resist being mischievous. He started to set other local mathematicians challenges involving solving equations of degree three. Of course, he won every time. The rumour of the existence of a solution then began to spread quietly.

In 1535, a Venetian scholar by the name of Niccolò Fontana Tartaglia was challenged by Del Fiore. Tartaglia was then aged thirty-five and had not yet published any scientific work. Del Fiore was thus unaware that he had just challenged someone who was going to become one of the best mathematicians of his generation. The two scholars exchanged a list of thirty questions for a stake of thirty banquets to be given by the loser to the winner. For several weeks, Tartaglia tore his hair over the problems of degree three sent to him by Del Fiore. But just a few days before the cut-off date, he managed to find the formula. He then solved the thirty problems in a few hours and won the challenge hands down.

The story might have stopped there, but here's what happened: Tartaglia in his turn refused to make his method public. This situation persisted for another four years.

It was then that the matter came to the attention of a mathematician and engineer from Milan by the name of Girolamo Cardano. His

Gallicized name, Jérôme Cardan, will doubtless ring a bell with mechanics buffs: he was, among other things, the inventor of Cardan joints, which in modern cars transmit the rotation of the engine to the wheels. Up to then, Cardano had been among those who thought the solution of equations of degree three was impossible. Intrigued by the challenge that Tartaglia had won, he then tried to befriend the latter. At the start of 1539, he had him sent eight problems to solve and asked him to pass on his method. Tartaglia refused point-blank. Cardano was annoyed, and then attempted an intimidating ploy by calling upon all the local algebraists to denounce their colleague's arrogance. Tartaglia did not give in.

Finally, Cardano achieved his ends with a ruse. He made it known to Tartaglia that Alfonso d'Avalos, the governor of Milan, wished to meet him. In Venice, Tartaglia was then in a precarious situation and needed a protector. He accepted the invitation to go to Milan, where the interview was arranged for 15 March 1539, in Cardano's house. Tartaglia waited in vain for the governor for three days. This time was enough for Cardano to overcome his guest's mistrust. After tireless negotiations, Tartaglia finally yielded, on the condition that Cardano swore never to publish his method. The oath was sworn and the formulae were delivered.

Back in Milan, Cardano set about dissecting the formulae. The method worked a treat, but there was still one thing he was missing – a proof. Up to that time, none of the mathematicians concerned had managed to give a rigorous proof that their formulae applied in all cases. It was to this task that Cardano applied himself over in the years that followed. He finally succeeded, and one of his pupils, Ludovico Ferrari, even managed to generalize the method to solve equations of degree four. However, bound by the oath they had sworn in Milan, the two mathematicians were unable to publish their results.

But Cardano did not let the matter drop. In 1542 he went with Ferrari to Bologna to meet Hannibale Della Nave, another former disciple of Scipione Del Ferro. Between the three of them, they managed to lay hands on Del Ferro's old notes, and determined that it was in fact he who had been the first to find the formulae. From that point on, Cardano decided that he was freed from his oath. In 1547 he published *Ars magna*, or *The Great Art*, a work that finally revealed to the world the method for solving equations of degree three. Tartaglia, who was furious, violently insulted Cardano and published his own version of the story. It was too late. In the eyes of the world, Cardano had become the one who had beaten degree three, and the method is known to this day as Cardano's formula.

However, a few details in the *Ars magna* gave rise to a certain scepticism among the algebraists of the period. In several cases, Cardano's formulae appeared to require the calculation of square roots of negative numbers. Reading through one equation, you could for example spot the occurrence of the root of −15, which, by definition, was assumed to be the number whose square is −15. But that was absolutely impossible by virtue of Brahmagupta's sign rule. The square of a positive number is positive, but the square of a negative number is also positive! For example, $(-2)^2 = -2 \times -2 = +4$. No number multiplied by itself can give −15. In short, the square roots that appeared in the calculation of these solutions simply did not exist. But – and here is the rub – in using these non-existent numbers as intermediate steps, Cardano's method nevertheless managed to come up with the right result. That was bizarre and intriguing.

It was another mathematician from Bologna, Rafael Bombelli, who studied this problem and suggested that the roots of negative quantities might well be a completely new type of number. These would be numbers that were neither positive nor negative, numbers of a strange and previously unheard-of nature whose existence no one had had any reason to suspect until then. Following the arrival of the zero and the negative numbers, the great family of numbers was once again about to become larger.

At the end of his life, Bombelli wrote his major work, *Algebra*, which he published in the year of his death in 1572. In it he returned to the discoveries of *Ars magna* and introduced those new creatures, which he called sophisticated numbers. Bombelli did for them what Brahmagupta had done in his day for the negative numbers. He listed the set of rules for calculations which governed the sophisticated numbers and in particular made their squares negative.

Bombelli's sophisticated numbers had a destiny similar to that of the negative numbers. They also generated their share of scepticism and incredulity. However, they too eventually became established, and their power revolutionized the world of mathematics. Among the converted sceptics in the 1630s was the French mathematician and philosopher René Descartes. It was he who gave these newcomers the name under which we still know them today: the imaginary numbers.

It took another two centuries for the imaginary numbers to be fully accepted by the whole of the mathematical community. From that point onwards they became indispensable in modern science. Beyond equations, these numbers have found multiple applications in the physical sciences, in particular in the study of all the wave-related phenomena that we find, for example, in electronics or quantum physics. Without them, numerous modern technological innovations would not have been possible.

However, unlike the negative numbers, the imaginary numbers remain little known outside scientific circles. They go against intuition, are difficult to conceive of, and do not represent simple physical phenomena. While the negative numbers could still be understood as a debt or a deficit, with the imaginary numbers you have to definitively stop thinking of numbers as quantities. It is impossible to give them a meaning that is applicable to day-to-day life, or to count apples or sheep with them.

The imaginary numbers slowly relieved mathematicians of their ultimate complexes. After all, if simply accepting the existence of negative square roots allows one to create a new type of numbers, why could one not go even further? Might it not be possible to add new numbers at will, provided one establishes their arithmetic properties? Might one not even be able to invent new algebraic structures that are totally independent from the classical numbers?

In the nineteenth century the last remaining prejudices about what numbers should be were abolished. From then onwards an algebraic structure became simply a mathematical construction consisting of elements (which one could call numbers in certain contexts but not always), and operations that can be performed on these elements (which one could call addition, multiplication, etc. in certain contexts, but not always).

This new freedom gave rise to a mighty explosion of creativity. New more or less abstract algebraic structures were discovered, studied and classified. Faced with the scale of the task, the mathematicians of Europe, then of the world, organized themselves, shared things and collaborated. To this day, algebraic research is extensively undertaken all around the globe, and numerous conjectures remain to be proved.

INVENTING YOUR OWN MATHEMATICAL THEORY

Have you ever dreamt of having a theorem to your name, like Pythagoras, Brahmagupta or al-Kashi? Great – I'll now demonstrate how you can create and study your own algebraic structure. You will need two ingredients for this: a list of elements and an operation that lets you combine them.

Take, for example, eight elements denoted by the following symbols: ♥, ♦, ♣, ♠, ♪, ♫, ▲ and ☼. You will also need a sign for your operation, so take, for example, ∗ and call it a 'Bombellification' in homage to the Italian scholar. To determine the result of the Bombellification of two elements, you now need to set up a table for this operation. Draw a table of eight rows and eight columns corresponding to your eight elements and fill it as you see fit by putting one of the elements in each cell.

That's it! Your theory is ready, all you have to do now is study it. Looking at the second row and the fourth column you can see, for example, that in Bombellifying ♦ by ♠, you obtain ☼. In other words, ♦ ∗ ♠=☼. You can even solve equations in your theory.

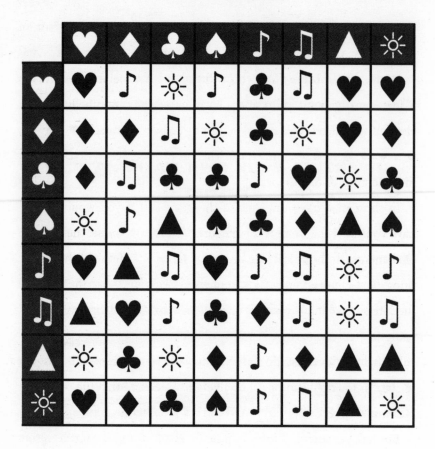

Look at this one:

FIND A NUMBER THAT GIVES ♬ WHEN YOU BOMBELLIFY IT BY ♣.

To find possible solutions, you just have to cast an eye on your table. You will see that there are two solutions: ♦ and ♪, since ♦∗♣ = ♬ and ♪∗♣ = ♬.

However, we need to be careful, because in your new theory, certain properties we are used to may become false. For example, the result may be different depending on the order in which you Bombellify two elements: ♥ ✳ ♦ = ♪ while ♦ ✳ ♥ = ♦. In this case, we say that the operation is not commutative.

With a little observation, you will nevertheless be able to discover a few more general properties. For example, by Bombellifying an element with itself, you always return to the element itself: ♥ ✳ ♥=♥, ♦ ✳ ♦=♦, ♣ ✳ ♣=♣, and so on. This result deserves to be called the first theorem of our theory.

In short, you have understood the principle. If you would like to have your own theorems, it's your turn to play. Of course, the number of elements can be as large as you wish, and even infinite, if that's your thing. You can define more complicated notation, as is the case for the integers, which do not all have their own symbol, but are written using the ten Indian figures. You will then be able to add calculation rules that will serve as axioms in your theory. For example it is possible to state in the definition of your algebraic structure that the operation is commutative.

But let's not deceive ourselves here; when you go about things in this way, there's little hope that your theory will pass to posterity. Not all mathematical models are of equal merit. Some are more useful and more important than others. When you create your operation table randomly, there's a strong chance that your model may be completely uninteresting. And if that were ever not the case, then it would be a good bet that another mathematician had already studied it before you.

Because, well, without any exaggeration, mathematics is a profession.

How do people recognize an interesting theory? Throughout history, two main criteria have guided mathematicians in their explorations. The first is usefulness, the second is beauty.

Usefulness is undoubtedly the most evident point. The primary reason for mathematics was to serve some purpose. Numbers are useful because they enable you to count and trade. Geometry makes it possible to measure the world. Algebra allows you to solve problems of everyday life.

Beauty, however, may seem a more fluid and less objective criterion. How can a mathematical theory be beautiful? This can be better understood in geometry, where certain figures can be appreciated visually like works of art. This is the case with the friezes of the Mesopotamians, the Platonic solids and the tilings of the Alhambra. But what about algebra? Can an algebraic structure really be beautiful?

I believed for a long time that the privilege of being touched by the elegance or the poetry of mathematics was something for specialists, for the privileged, something that only enlightened enthusiasts – those who have spent enough time studying, dissecting and digesting the theories in their most minute detail, those who have developed a profound and mature intimacy with the abstract concepts – could appreciate. I was mistaken, and I have noticed on many occasions that this feeling of elegance can touch complete novices, and even very young children.

I observed one of the most striking examples of this one day when I was leading research workshops with a class of primary school children. The children were about seven years old. They had to manipulate triangles, squares, rectangles, pentagons, hexagons and various other shapes, which they were tasked to sort according to criteria of their choice. It had become obvious that for each of the figures we could count the number of its sides and also the number of its vertices. Triangles have 3 sides and 3 vertices, squares and rectangles have 4 sides and 4 vertices, and so on. In drawing up this list, the children had rapidly come up with a theorem: a polygon always has as many sides as it has vertices.

The following week, by way of a challenge for them, we brought in more quirky figures, including one of the following shape:

The question then was: how many sides and how many vertices does it have? And the majority of the class replied 4 sides and 3 vertices. The inverted corner at the bottom of the figure does not come to a peak. It isn't pointed. This figure can't be rolled. It is a hollow rather than a bump. In short, this re-entrant angle did not fit in with their prior conception of a vertex. Asking them to call this point a vertex was asking them to give the same name to different things. What an idea! Discussions ensued. Not all the children were in agreement about the status of this new point. Should it be given another name? Should it be ignored completely? There were arguments for and arguments against, but overall, none seemed to convince the majority.

Then, suddenly, one child remembered the theorem. If that is not a vertex, then we can no longer say that every polygon always has as many sides as it has vertices. To my great astonishment, it was this argument that instantly swung the class. In a few seconds, everyone agreed: this point should be called a vertex. The theorem had to be saved, even at the cost of our prejudices. It would have been too great a shame for such a simple and clear statement to have exceptions. This was the most precocious occurrence of the feeling of mathematical elegance in young children that I have ever witnessed.

Exceptions are not beautiful. Exceptions are sickening. The simpler a statement is, and the greater its significance, the more it gives us the impression that we are touching something deep with the tips of our fingers. Beauty in mathematics can take several forms which all manifest themselves through that disturbing relationship between the complexity of the objects and the simplicity of their formulation. A beautiful theory is an economic theory, with no waste, no arbitrary exceptions or pointless distinctions. It's a theory that says a lot in a little, that settles the essential things in a few words, that is instantly immaculate.

While the example of polygons remains elementary, the impression of elegance only increases as theories grow and at the same time retain an order that reduces to a few simple rules. It is even more disturbing when a new theory that one might have thought was more complicated than the old one turns out in reality to be better adapted and more harmonious. The imaginary numbers are a perfect illustration of this.

Do you remember equations of degree two? According to al-Khwārizmī's method, it was possible for these equations to have two solutions, but it was also possible for them to have just one, or even for them to have none at all. That is valid provided you only consider solutions that do not involve imaginary numbers. If you take the latter into account, the rule becomes far simpler: all equations of degree two have two solutions! When al-Khwārizmī claimed that an equation did not have a solution, it was simply because he was stuck in a set of numbers that was too narrow. These two solutions were imaginary.

But there was better to come. Thanks to the imaginary numbers, all equations of degree three have three solutions, all equations of degree four have four solutions, and so on. In short, the rule is as follows: The number of solutions of an equation is equal to its degree. This result was conjectured in the eighteenth century before being proved at the start of the nineteenth by the German mathematician Carl Friedrich Gauss. Today it is called the fundamental theorem of algebra.

More than one thousand years after al-Khwārizmī's treatise, after all the setbacks at degree three, after the difficulties in conceiving of equations beyond degree four without geometric representation, who would have thought that it would all end with a simple rule of twelve words? The number of solutions of an equation is equal to its degree.

Such is the wonder of the imaginary numbers. And equations are not the only things to benefit from them. In the imaginary world, numerous theorems can suddenly be stated with a breath-taking concision and elegance. All the pieces of the mathematical jigsaw seem to fit together wonderfully with them. Bombelli probably did not suspect that in legitimizing his 'sophisticated numbers', he was timidly opening the door to a true paradise for generations of mathematicians.

In the new algebraic structures that bloomed in the nineteenth century, mathematicians sought properties of the same type: general rules, symmetries, analogies, results that followed on from and complemented one another to perfection. The small theory we invented earlier falls far short of meeting these criteria, which would make it interesting. It is perfectly random, and almost everything in it is a special case.

There are no great general rules concerning either the equations, or the properties of its operation; that's too bad.

One of the great names of modern algebra was the Frenchman Évariste Galois, a precocious genius who died in 1832 at the age of twenty following a duel, but who, in his short existence, even found time to make his contribution to the history of equations. Galois was able to prove that from degree five onwards, the solutions of certain equations could no longer be calculated using formulae similar to those of al-Khwārizmī or Cardano, which only use the four operations, powers and roots. For his particularly brilliant proof, he created custom-built new algebraic structures, which are still studied today under the name of Galois groups.

But the person who was perhaps the most prolific in the art of deducing major algebraic results from a restricted number of elementary axioms was the German mathematician Emmy Noether. From 1907 to her death in 1935, Noether published almost fifty articles on algebra, some of which revolutionized the discipline through her choice of algebraic structures and the theorems that she deduced from them. She mainly studied what we now call rings, fields and algebras,* that is to say structures that have, respectively, three, four and five operations interlinked by well-chosen properties.

Algebra then entered realms of abstraction in the face of which this modest book must give way to university courses and academic works.

---

* The word 'algebra' denotes both the discipline as a whole and a particular type of algebraic structure.

12

A LANGUAGE FOR MATHEMATICS

Sixteenth-century Europe simmered. The Renaissance had spilled over from Italy and was inundating the whole continent. Innovations came one after the other and discoveries proliferated. To the west, across the Atlantic, Spanish ships had just found a new world. And while ever-greater numbers of explorers were setting off in search of distant lands, the humanist intellectuals, in their libraries, were going back in time and rediscovering the great texts of antiquity. On the religious front too, traditions were being overturned. The Protestant Reformation led by Martin Luther and John Calvin was experiencing a growing success, and the French Wars of Religion would rage in the second half of the century.

The propagation of these new ideas was broadly supported by the arrival of a brand-new invention perfected in the 1450s by the German Johannes Gutenberg: the press with movable type. Thanks to this process, it became possible to print many copies of a book very quickly and to circulate it on a large scale. In 1482, Euclid's *Elements* was the first mathematical work to pass through the presses in Venice; the procedure proved a dazzling success. At the start of the sixteenth century, several hundred towns had their own press and tens of thousands of works had already been printed.

The sciences played an active part in these upheavals. In 1543, the Polish astronomer Nicolaus Copernicus published his *De revolutionibus orbium coelestium*, or *On the Revolutions of the Celestial Spheres*. This came as a bolt from the blue. Casually discarding the astronomical system of Ptolemy, Copernicus asserted that it is the Earth that revolves around the sun, and not the other way round. In the years that followed, Giordano Bruno, Johannes Kepler and also Galileo Galilei followed in his footsteps, establishing heliocentrism – the theory of the sun at the centre – as the new reference model for cosmology. This revolution could not fail to bring down the wrath of the Catholic Church upon the scholars who supported it; after having encouraged the expansion of the sciences for some time, the Church was caught off its guard when they rejected its dogmas. While Copernicus had had the presence of mind not to publish his works until shortly before his death, Bruno was burnt to death in public in Rome and in 1633 Galileo was forced to recant in the Court of the Inquisition. The legend has it that as he left the courtroom, the Italian murmured between his lips four words that have since become famous: '*E pur si muove!*' ('And yet it moves!')

Mathematics followed the movement of ideas, and descended gradually over the great kingdoms of Western Europe, and in particular on France.

Of course, mathematics had been practised on French territory before that period. The Gauls had their numeration system in base twenty, and the French *quatre-vingts* (meaning four twenties) for eighty is undoubtedly a relic of this. The Romans who occupied Gaul may not have been great mathematicians, but they had a sufficient mastery of figures to administer their gigantic empire efficiently. The same applied to the Franks, the Merovingians, the Carolingians and the Capetians who succeeded one another through the Middle Ages. However, France had never had any top-ranking mathematicians. No theorems or major results had ever been discovered in France that had not already been discovered elsewhere in the world.

Just as the mathematicians descended on France, it's time for me to take to the road: in the direction of the Vendée. Today, I have a rendezvous in the west of the country with the first great French mathematician of the Renaissance: François Viète (1540–1603).

The village of Foussais-Payré, 12 kilometres from Fontenay-le-Comte, is laden with history. The first traces of occupation of the site go back to the Gallo-Roman period, but it was in the Renaissance that the village enjoyed a time of great prosperity. Artisans and traders settled here in large numbers and their businesses flourished. Trade in wool, linen and leather made them famous in all four corners of the kingdom. To this day, numerous buildings from that period remain remarkably well preserved. For its 1,000 inhabitants, the village has no less than four buildings that are classified as historic monuments, and many other ancient dwellings.

To the north of the village stands the place called La Bigotière, the former tenant farm that François Viète inherited from his father and that earned him his title of Seigneur de La Bigotière. In the main street is the Auberge Sainte-Catherine, a former property of the family where Viète liked to spend time in his adolescence. There is something very moving about entering the walls within which France's first great mathematician grew up. The young François undoubtedly spent many winter evenings beside this enormous fireplace which takes centre stage in the main room, now converted into a restaurant. Were his mathematical thoughts first kindled by the heat of this fire?

Viète did not spend all his life in Foussais-Payré. After studying law in Poitiers, he travelled to Lyon, where he was presented to King Charles IX, then he spent some time in La Rochelle before moving to Paris.

The French Wars of Religion were then at their height. François's family was itself divided over the question. His father, Étienne Viète, converted to Protestantism, while his two uncles remained Catholics. François kept aloof from these debates and never revealed his deep convictions. He was in turn a lawyer for major Protestant families and a high dignitary of the kingdom. His hedging meant that he was not always well looked upon, and he went through several periods of disgrace. On Saint Bartholomew's Day in 1572 he was in Paris, but escaped the massacre that slaughtered many thousands of Huguenots. Not everyone was so fortunate. Pierre de La Ramée, who had been the first to introduce mathematics at the University of Paris, and whose works had a strong influence on Viète, was killed on 26 August.

In parallel with his official responsibilities, Viète practised mathematics as an amateur enthusiast. Of course he knew Euclid, Archimedes and the scholars of antiquity whose texts were rediscovered by the Renaissance. He was also interested in the Italian scholars, and was one of the first to read Bombelli's *Algebra*, whose publication had gone more or less unnoticed at that time. However, Viète continued to side with the sceptics where the introduction of sophisticated numbers was concerned. Throughout his life, he published his mathematical works at his own expense to offer them to anyone he deemed worthy of reading them. He was interested in astronomy, trigonometry, and also cryptography.

In 1591 Viète published what was to become his main work: *In artem analyticem isagoge*, or *Introduction to the Art of Analysis*, which was often simply called the *Isagoge*. Curiously, the *Isagoge* did not go down in history because of the theorems or the mathematical proofs that he developed there, but for the way in which these results were formulated. Viète was to become the main instigator of the new algebra that rose, in a matter of decades, to become a whole new mathematical language.

In order to understand his approach, we have to delve back into the mathematical works of earlier times. While the geometric theorems of Euclid or the algebraic methods of al-Khwārizmī are still very useful nowadays, the way in which they are expressed has been radically transformed. Ancient scholars did not have a specific language for writing mathematics. All the symbols with which we are so familiar, such as those used for the four elementary operations: '+', '−', '×' and '÷', were only invented in the Renaissance. For almost five millennia, from the Mesopotamians to the Arabs, by way of the Greeks, the Chinese and the Indians, mathematical formulae were squatters in the everyday vocabulary of the languages in which they were written.

The books of al-Khwārizmī and the algebraists of Baghdad are thus entirely written in Arabic, without any symbolism. In these works, certain arguments could then extend over several pages where even a few lines would suffice nowadays. Do you recall the following equation of degree two presented in the *al-jabr*:

THE SQUARE OF A NUMBER PLUS 21 IS EQUAL TO 10 TIMES THAT NUMBER.

Here is how al-Khwārizmī detailed his solution.

THE SQUARES AND THE NUMBERS ARE EQUAL TO THE ROOTS; FOR EXAMPLE, 'A SQUARE AND TWENTY-ONE IN NUMBERS ARE EQUAL TO TEN ROOTS OF THE SAME SQUARE'. THAT IS TO SAY, WHAT MUST BE THE QUANTITY OF A SQUARE WHICH, WHEN TWENTY-ONE DIRHAMS ARE ADDED TO IT, BECOMES EQUAL TO THE EQUIVALENT OF TEN ROOTS OF THAT SQUARE? SOLUTION: TAKE HALF OF THE NUMBER OF ROOTS; HALF IS FIVE. MULTIPLY IT BY ITSELF; THE PRODUCT IS TWENTY-FIVE. TAKE AWAY FROM THIS TWENTY-ONE WHICH IS ASSOCIATED WITH THE SQUARE; THE REMAINDER IS FOUR. EXTRACT ITS ROOT; IT'S TWO. TAKE THAT AWAY FROM HALF OF THE ROOTS, WHICH IS FIVE; THAT LEAVES THREE. THAT IS THE ROOT OF THE SQUARE THAT YOU ARE LOOKING FOR AND THE SQUARE IS NINE. YOU COULD ALSO ADD THE ROOT TO HALF OF THE ROOTS; THE SUM IS SEVEN; THAT IS THE ROOT OF THE SQUARE WHICH YOU ARE LOOKING FOR AND THE SQUARE ITSELF IS EQUAL TO FORTY-NINE.

Nowadays, such a text is very laborious to read, even for students who have a perfect grasp of the method in question. The text leads to two solutions: 3 and 7 with squares 9 and 49.

Rhetorical algebra, as it would later be called, is not only very long-winded to write, but it also suffers from certain ambiguities of the language that can give a single phrase several interpretations. As arguments and proofs became more complicated, this mode of writing progressively proved to be a nuisance to handle.

These difficulties are occasionally compounded by those that some mathematicians bring upon themselves. We often find mathematics written in verse. This practice is often the relic of an oral tradition in which learning by rote is assisted by the poetic form. When Tartaglia passed on his method for solving equations of degree three to Cardano, he wrote in Italian and in alexandrines.

Evidently, the proof lost in clarity what it gained in poetry, and we may well suspect that Tartaglia, who we know was reluctant to divulge his proof, deliberately clouded its comprehension. Here is an extract translated into English.

> **WHEN THE CUBE AND ITS THINGS NEAR**
>
> **ADD TO A NEW NUMBER, DISCRETE,**
>
> **DETERMINE TWO NEW NUMBERS DIFFERENT**
>
> **BY THAT ONE; THIS FEAT**
>
> **WILL BE KEPT AS A RULE**
>
> **THEIR PRODUCT ALWAYS EQUAL, THE SAME,**
>
> **TO THE CUBE OF A THIRD**
>
> **OF THE NUMBER OF THINGS NAMED.**
>
> **THEN, GENERALLY SPEAKING,**
>
> **THE REMAINING AMOUNT**
>
> **OF THE CUBE ROOTS SUBTRACTED**
>
> **WILL BE OUR DESIRED COUNT.**

This is pretty obscure, isn't it? What Tartaglia calls 'things' is precisely the number sought, the unknown. The presence of cubes in this text is a clear indication that we are dealing with an equation of degree three. Once he was in possession of the poem, Cardano himself had the greatest difficulty deciphering it.

In order to cope with this growing complexity, mathematicians gradually began to simplify the algebraic language. This process began in the Muslim Occident in the last centuries of the Middle Ages, but it was above all in Europe between the fifteenth and sixteenth centuries that the movement spread to its full extent.

In the early stages, new words specific to mathematics made their appearance. Thus, in the mid-sixteenth century, the Welsh mathematician Robert Recorde proposed a nomenclature for certain powers of an unknown number, based on a system of prefixes that made it possible to multiply the powers as far as one wished. The square of the unknown was, for example, called zenzike, its sixth power zenzicubike and its eighth power zenzizenzizenzike!

Then, little by little, brand-new symbols, which seem quite familiar to us today, began to break out all over and at random.

In around 1460, the German Johannes Widmann was the first to use the signs + and − to denote addition and subtraction. In the mid-fifteenth century, Tartaglia, whom we have met, was one of the first to use parentheses ( ) in calculations. In 1557, Robert Recorde first used the sign = to denote equality. In 1608, the Dutchman Rudolph Snellius used a comma as a separatrix between the integer part and the decimal part of a number. The signs < and > – 'is less than' and 'is greater than' – to indicate the numerical inferiority or superiority of two numbers were introduced by the Englishman Thomas Harriot in his book *Artis Analyticae Praxis,* published after his death, in 1631.

Also in 1631, the Englishman William Oughtred used the cross × to denote multiplication, and in 1647 he became the first to use the Greek letter π to denote Archimedes' famous ratio. The German Johann Rahn was the first to use the obelus ÷ for division in 1659. In 1525 another German, Christoff Rudolff, denoted the square root by the sign $\sqrt{\ }$, to which René Descartes added a horizontal bar in 1647: $\sqrt{\ \ }$.

Of course, all that did not happen in an orderly, linear manner. A host of other symbols were born and died in that same period. Some of them were only used once. Others were developed and vied for survival. Often, several decades elapsed between the first use of a sign and its definitive adoption by the mathematical community as a whole. Thus, a century after their introduction, the signs + and − had not been fully adopted, and many mathematicians were still using the letters P and M (being the initial letters of the Latin words *plus* and *minus*) to denote addition and subtraction.

And where was Viète in all this? The French scholar was one of the catalysts of this vast movement. In the *Isagoge*, he initiated a vast programme to modernize algebra and laid the keystone for this by introducing literal calculus, that is to say arithmetic with the letters of the alphabet. His proposal was as simple as it was puzzling: to refer to the unknowns in equations by vowels and to the known numbers by consonants.

This split into vowels and consonants was, however, soon rapidly abandoned in favour of a slightly different suggestion by René Descartes: namely that the first letters of the alphabet (a, b, c . . .) should denote known quantities and the last letters (x, y and z) should stand for the unknowns. This is the convention that most mathematicians still use today, and the letter 'x' has even entered everyday language as a symbol for the unknown and the mysterious.

To understand how algebra was transformed by this new language, we recall the following equation:

FIND A NUMBER THAT WHEN MULTIPLIED BY 5 GIVES 30.

Using the new symbolism, this equation could then be written using a handful of signs: $5 \times x = 30$.

Much shorter; do you also remember that this equation was only a special case from a much broader class? :

**FIND A NUMBER THAT WHEN MULTIPLIED BY A CERTAIN QUAN-
TITY 1 GIVES A QUANTITY 2.**

This equation could then be written as $a \times x = b$.

Since the numbers signalled by a and b come from the start of the alphabet, we know that these are known quantities from which we seek to calculate x. And, as we saw, equations of this type are solved by dividing the second known quantity by the first – in other words: $x = b \div a$.

From then on, mathematicians set about drawing up lists of cases and establishing rules for manipulating literal equations. Algebra was gradually transformed into a kind of game in which the authorized turns are determined by these arithmetic rules. For example, let's look again at the solution of our equation. In going from $a \times x = b$ to $x = b \div a$, the letter a moves from the left to the right of the equals sign and its operation is transformed from multiplication into division. This then is an authorized rule: any quantity that is multiplied can be moved to the other side of the equality, where it becomes a divisor. Similar rules are used to deal with additions and subtractions and for transforming powers. The aim of the game remains the same: to determine the value of the unknown x.

This game with symbols was so effective that algebra rapidly gained its independence from geometry. It was no longer necessary to interpret multiplications as rectangles, or to have proofs in the form of jigsaws. The x, the y and the z took over. Better than that, the dazzling efficiency of literal calculus overturned the power relationship, and soon geometry found itself dependent upon algebraic proofs.

It was René Descartes who triggered this reversal by introducing a simple and powerful way of 'algebrizing' problems in geometry using a system of axes and coordinates.

CARTESIAN COORDINATES

Descartes' idea was as elementary as it was brilliant: to draw two graduated straight lines in the plane, one horizontal and the other vertical, in order to identify each geometric point by its coordinates on these two axes. For example, take the point A:

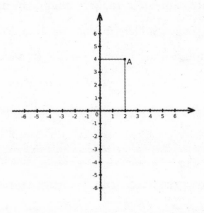

The point A lies above graduation 2 on the horizontal axis and at the level of graduation 4 on the vertical axis. Its coordinates are thus 2 and 4. Using this procedure, it becomes possible to represent any geometric point by two numbers, and conversely to associate a point with any pair of numbers.

From their beginnings, geometry and numbers have always been closely related, but with Cartesian coordinates the two disciplines came together. Every problem in geometry could then be interpreted algebraically, and every problem in algebra could be represented geometrically.

Let us consider, for example, the following equation of degree one: x + 2 = y. This is an equation in two unknowns: we seek x and y. You can see, for example, that x = 2 and y = 4 form a solution, since 2 + 2 = 4. You may then note that the numbers 2 and 4 are precisely the coordinates of the point A. This solution can thus be represented geometrically by that point.

In fact, the equation x + 2 = y has infinitely many solutions. These include, for example x = 0 and y = 2, or again x = 1 and y = 3. For each possible value of x, one can find the corresponding value of y by adding 2 to it. We can then insert all the points corresponding to these solutions on our diagram. Here is what we obtain:

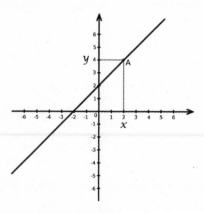

The solutions are all perfectly aligned to form a straight diagonal line – there isn't one that goes astray. In Descartes' world, this line is then the geometrical representation of the equation, just as the equation is the algebraic representation of the line. The two objects blend together and mathematicians now commonly speak about the line 'x + 2 = y'. We are giving the same name to different things. With the introduction of Cartesian coordinates, algebra and geometry were well on the way to becoming one and the same discipline.

This correspondence gave rise to a whole dictionary for translating objects in the language of geometry into the language of algebra and vice versa. For example, what we call the 'middle' in geometry is known as the 'average' or 'mean' in algebra. Let us return to our point A with coordinates 2 and 4 and add a point B, with coordinates 4 and −6. To find the middle of the segment linking A and B, it now suffices to take the average of the coordinates. The first coordinate of A is 2 and that of B is 4, so we can deduce from this that the first coordinate of the middle is equal to the average of these two numbers: $(2 + 4) \div 2 = 3$. Doing the same thing on the vertical axis, we find $(4 + -6) \div 2 = -1$. The coordinates of the midpoint are therefore 3 and −1. We can check that this works by drawing the diagram:

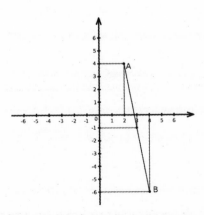

In this bilingual dictionary from geometry to algebra, a circle becomes an example of an equation of degree two, the point of intersection of two curves is given by a system of equations, while Pythagoras' theorem, trigonometric constructions and carving up into jigsaw puzzles are transformed into various literal formulae.

In other words, it was no longer necessary to draw figures to do geometry; algebraic calculations had replaced them, and were much faster and more practical.

In the centuries that followed, Cartesian coordinates recorded numerous successes. One of their greatest triumphs was undoubtedly the resolution of a conjecture that had resisted mathematicians since antiquity, namely the squaring of the circle.

Using a ruler and a compass, is it possible to draw a square with the same area as a given circle? You will remember that, more than three thousand years ago, the scribe Ahmes had already been tearing his hair over this question. After him, the Chinese and the Greeks had tried their hand at it with no greater success, and over the centuries the problem had become one of the greatest conjectures of mathematics.

Through Cartesian coordinates, straight lines constructed with a ruler were transformed into equations of degree one, while circles drawn with a compass became equations of degree two. From an algebraic point of view, the squaring of the circle could thus be stated as follows: is it possible to find a series of equations of degree one or two for which the number π is a solution? This new formulation reignited research, but even when posed in this way, the question remained complicated.

It was the German mathematician Ferdinand von Lindemann who finally put an end to the suspense in 1882. No, the number π is not a solution of equations of degree 1 or 2, and it is thus impossible to square the circle. So it was that, having resisted the assaults of mathematicians for so long, the problem, which to this day is still referred to as a conjecture, was resolved.

Cartesian coordinates are easily generalizable to geometry in three dimensions. There, every point is then identified by three coordinates and the algebraic procedures can be applied in the same way.

Things grow more subtle when we move to the fourth dimension. In geometry, it is impossible to represent a figure in 4D because the whole of our physical world is itself only in 3D. In algebra, on the other hand, there is no such difficulty: a point in the fourth dimension is simply a list of four numbers, and all the algebraic methods apply there naturally. If we consider, for example, the points A and B with coordinates 1, 2, 3 and 4 and 5, 6, 7 and 8, respectively, we can calmly use the average of these numbers to assert that their centre is the point with coordinates 3, 4, 5 and 6. Geometry in dimension four found particular application in the twentieth century in Albert Einstein's theory of relativity, in which the fourth coordinate is used to model time.

And one could go on for ever in the same way. A list of five numbers is a point in dimension 5. Add a sixth number and you are in dimension 6. There is no limit to this process. A list of one thousand numbers is a point in a space of dimension 1,000.

At this level, the analogy may seem like a simple game with language, which can bring a smile, but has no real utility. But make no mistake about it, this correspondence has many applications, particularly in statistics, where the object is precisely to study long lists of numerical data.

For example, if you are studying the demographic data of a population, you may wish to quantify the extent to which certain characteristics such as the height, weight, or eating habits of a group of individuals fluctuate around the mean. The geometrical interpretation of this question involves calculating the distance between two points, the first representing the list of data concerning each individual, the second representing the average list. The number of coordinates is equal to the number of individuals in the group. The calculation involves the use of right-angled triangles in which one can apply Pythagoras' theorem. A statistician calculating the typical deviation in a group of one thousand individuals is thus using, often without knowing it, Pythagoras' theorem in a space of dimension 1,000. This method is also used in evolutionary biology to calculate the genetic difference between two animal populations. When we use formulae originating in geometry to measure the distance between their genomes encoded in the form of a list of numbers, it becomes possible to establish the relative proximity of different species and to gradually deduce from that the genealogical tree of the living animal.

We can even push the exploration up to infinite lists of numbers, that is to say to points in a space of infinite dimension. In fact, we have seen this already in numerical sequences such as the Fibonacci sequence. When he studied his rabbits, the Italian mathematician was unsuspectingly doing infinite-dimensional geometry. In particular, it was this geometric interpretation that enabled the mathematicians of the eighteenth century to establish the subtle link between the Fibonacci sequence and the golden ratio with the greatest possible clarity.

13

THE WORLD'S ALPHABET

PHILOSOPHY [NATURE] IS WRITTEN IN THAT GREAT BOOK WHICH EVER IS BEFORE OUR EYES – I MEAN THE UNIVERSE – BUT WE CANNOT UNDERSTAND IT IF WE DO NOT FIRST LEARN THE LANGUAGE AND GRASP THE SYMBOLS IN WHICH IT IS WRITTEN. THE BOOK IS WRITTEN IN MATHEMATICAL LANGUAGE, AND THE SYMBOLS ARE TRIANGLES, CIRCLES AND OTHER GEOMETRICAL FIGURES, WITHOUT WHOSE HELP IT IS IMPOSSIBLE TO COMPREHEND A SINGLE WORD OF IT; WITHOUT WHICH ONE WANDERS IN VAIN THROUGH A DARK LABYRINTH.

This paragraph, which is one of the most famous in the history of the sciences, was written in 1623 by Galileo in a work entitled *Il Saggiatore*, or *The Assayer*.

Galileo was incontestably one of the most prolific and innovative scientists of all time. The Italian scholar is generally considered to be the founder of the modern physical sciences. His curriculum vitae is impressive, to say the least. He invented the astronomical telescope. He discovered the rings of Saturn, sunspots, the phases of Venus and the four main satellites of Jupiter. He was one of the most influential defenders of Copernicus's heliocentrism, stated the principle of the relativity of motion that now carries his name, and was the first to study falling bodies experimentally.

The Assayer bears witness to the very strong connection between mathematics and physical sciences that was being forged at that time. Galileo was one of the first to engage with this shift in relations. In fact, it has to be said that he had been to the right school, because at the age of nineteen he was initiated into mathematics by one of Tartaglia's students, Ostilio Ricci. He was followed by generations of scientists for whom algebra and geometry would beyond doubt become the language in which the whole world expressed itself.

Let us be quite clear about the nature of that nascent relationship between mathematics and physics. Of course, as we have already observed many times in this story, mathematics has always been used to study and understand the world. However, what happened in the seventeenth century was radically new. Up to then, mathematical modelling had remained at the stage of man-made constructions: imitating reality, but not determined by it. When the Mesopotamian surveyors used geometry to measure a rectangular field, that field had been laid out by men. The rectangle did not belong to nature before the farmer put it there. Similarly, when the geographers triangulated a region in order to map it, the triangles they considered were purely artificial.

The desire to 'mathematize' the world that already existed before man is another challenge. It's true that a few scholars tried their hand at this in antiquity. This was the case for Plato, who, you will recall, associated the five regular polyhedra with the four elements and with the cosmos. The Pythagoreans themselves were particularly partial to this type of interpretation, but it must be acknowledged that most often their theories had nothing serious about them. Based on purely metaphysical considerations, and never tested experimentally, nearly all of them turned out in the end to be false.

What the scholars of the seventeenth century came to understand was that nature itself, down to its most minute details, is controlled by precise mathematical laws, and that light can be cast upon these laws through experiments. One of the most brilliant revelations of that period was without a doubt the universal law of gravitation, discovered by Isaac Newton.

In his *Philosophiae naturalis principia mathematica*, or the *Mathematical Principles of Natural Philosophy*, the English scholar was the first to understand that the falling of bodies to the Earth and the rotation of celestial bodies in the heavens can be explained by one and the same phenomenon. All objects in the Universe are attracted to one another. This force is practically undetectable for small objects, but becomes significant in the case of planets or stars. The Earth attracts objects, which is the reason why objects fall. The Earth also attracts the Moon, and, in a certain sense, the Moon also falls. But since the Earth is round and the Moon is moving at very high speed, the Moon keeps falling alongside the Earth, and that is what keeps it in orbit! It is this same principle that causes the planets to go round the sun.

Newton did not simply state this law of attraction. He also specified the intensity of force by which objects attract one another. And he specified it in a mathematical formula. Any two bodies are attracted by a force proportional to the product of their masses divided by the square of their distance. This was expressed in Viète's literal calculus as follows:

$$F = G \times \frac{m_1 \times m_2}{d^2}$$

In this formula, the letter F denotes the intensity of the force, m_1 and m_2 are the respective masses of the two objects whose attraction is being studied, and d is the distance between them. The number G is a fixed constant equal to 0.0000000000667. Its very small value indicates that the force is imperceptible for small objects, and that it takes the enormous masses of planets and stars for gravitation to make itself felt. But consider this: whenever you pick up an object, you are demonstrating that your muscular force is greater than the force of attraction of the entire Earth.

Once the formula was established, the physical problems turned into mathematical problems. It thus became possible to calculate the trajectories of celestial bodies, and in particular to predict their future evolution. To find the date of the next eclipse comes down to determining the value of an unknown in an algebraic equation.

Many successes were recorded in the decades following the discovery of Newton's formula. Universal gravitation led to the assertion that the Earth must be slightly flattened at the poles, which was indeed confirmed by the geometers who measured the Meridian by triangulation. However, one of the most spectacular successes of the Newtonian theory was the calculation of the return of Halley's Comet.

Since antiquity, scholars had observed and recorded the apparently random appearance of comets in the heavens. These phenomena were explained by two opposing schools of thought. Aristotelians considered the comets to be atmospheric phenomena, and hence relatively close to the Earth, while Pythagoreans saw them as kinds of planets, meaning much more distant objects. When Newton published his *Principia mathematica*, the polemic had not yet been resolved, and the scholars of the two schools were still arguing about the subject.

One way of proving that comets are distant celestial bodies orbiting the sun would have been to find a certain periodicity for them: an object that is in orbit must revisit the same point at regular intervals. Sadly, at the beginning of the eighteenth century, no regularity of this type had yet been detected. Then, in 1707, a British astronomer and friend of Newton by the name of Edmund Halley announced that he might have found something.

In 1682, Halley had observed a comet that he had not found extraordinary at first. However, in the previous year, the astronomer had visited France, where he had met Cassini I at the Paris Observatory. The latter had brought up the subject of the hypothesis of the periodic return of comets. Halley then delved into the astronomical archives, where two other passages of comets eventually attracted his attention, the first in 1531, and the second in 1607. The comets of 1531, 1607 and 1682 formed two nearly identical intervals of 76 and 75 years. What if they were all the same comet? Halley took a gamble and announced that the comet would return in 1758.

This meant fifty-one years of suspense. The wait was unbearable and hair-raising. Other scholars took advantage of it to refine Halley's prediction. It was, in particular, suggested that the gravitational attraction of the two giant planets, Jupiter and Saturn, might actually modify the trajectory of the comet slightly. In 1757, the astronomer Jérôme Lalande and the mathematician Nicole-Reine Lepaute embarked upon calculations based on a model developed by Alexis Clairaut using Newton's equations. The calculations were long and detailed, and it took several months for the three scholars to finally predict a passage of the comet at its closest to the sun in April 1759, with a possible margin of error of one month.

And then the incredible happened. The comet made the rendezvous and the whole world saw it mark out the triumph of Newton and Halley in the heavens. It passed alongside the sun on 13 March, within the interval calculated by Clairaut, Lalande and Lepaute. Unfortunately, Halley did not live long enough to witness the return of the comet to which he gave his name, but the theory of gravitation and, through that, the mathematization of physics had just given a striking proof of their incredible power.

It is an irony of history that in *The Assayer*, apart from his discourse on the mathematization of the world, Galileo supported the thesis of atmospheric comets. His book was actually a response to the mathematician Orazio Grassi, who had defended the opposing point of view several years earlier. Galileo's fame and the highly polemicized tone of his book made it a best-seller of the period, but neither fame nor success makes the truth, and Grassi's response to Galileo might have been: '*E pur si muove . . .* '

Aside from Galileo's mistake, this anecdote provides a superb illustration of the robustness of the scientific process that was being put in place in that period. The conclusions of the scientific method do not depend on the prior opinion of the scholar who practises it, even when that scholar is Galileo. Facts are stubborn. The real nature of comets, like that of all objects of the physical world, is independent of how men conceive of it. When, in antiquity, a famous scholar was mistaken, a whole set of disciples would follow him, mostly without demur, as authority prevailed over reasoning. Several centuries were often too few to dislodge a received idea that could have been debunked by a simple experiment. The fact that Galileo's mistake was detected within a few decades, on the other hand, was an indication of a very healthy scientific environment.

It is one thing to predict the trajectory of a comet that has already been seen, but quite another thing to calculate that of a heavenly body that is a complete unknown. The discovery of Neptune in the nineteenth century must also count as one of the great successes of mathematics in astronomy. The eighth and last planet of the solar system is the only one that was not discovered by observations but by calculations. This feat was achieved by the French astronomer and mathematician Urbain Le Verrier.

From the end of the eighteenth century onwards, several astronomers had remarked upon irregularities in the trajectory of Uranus, which was then the last known planet. The latter did not follow exactly the trajectory that the universal law of gravitation predicted for it. There could only be two explanations: either Newton's theory was wrong, or another, as yet unknown, heavenly body was responsible for these perturbations. Based on the observed trajectory of Uranus, Le Verrier set about calculating the position of this hypothetical new planet. It took him two years of relentless work to obtain a result.

Then came the moment of truth. On the night of 23–24 September 1846, the German astronomer Johann Gottfried Galle pointed his telescope in the direction that Le Verrier had communicated to him, looked through the eyepiece and saw it. What he actually saw was a small bluish spot lost in the depths of the night sky. At a distance of more than 4 billion kilometres from the Earth, the planet was indeed there.

One can but imagine the feelings of astonishment and intoxication, the impressions of the power of the universe, and the fathomless depths of emotion that must have overcome Urbain Le Verrier that day, with the realization that his sharpened pencil and the strength of his equations had enabled him to embrace, capture, and almost to regulate the titanic dance of the planets around the sun. Through mathematics, planets – the celestial monsters, the gods of times gone by – suddenly found themselves tamed, controlled, docile and purring to the touch of algebra. One can but imagine the state of intense exaltation that overcame the astronomical community across the world in the following days, and which to this day sends a shiver down the spine of every amateur astronomer who points a telescope towards Neptune.

The life of a scientific theory has its phases, or paradigm shifts. First there is the time for hypotheses, hesitations, errors and the gradual and sometimes tenuous construction of ideas. Then comes the time for confirmation, the time for the experiments that validate or invalidate the equations, and like implacable judges yield a definitive confirmation or rejection. Then there is the phase in which the theory takes off and claims its independence. This is when the theory grows so self-assured that it deigns to speak to the world without having to look it in the eyes, the time when equations may precede experiments and predict an as yet unobserved, unexpected or even unhoped-for phenomenon, the time when theory passes from the discovery to exploration, when it becomes the ally, almost the colleague, of the scholars who created it. Then, the theory is mature, and this is the time for Halley's comet and Neptune. It is also the time for Einstein's eclipses, which on 29 May 1919 saw the triumph of relativity, the time for the Higgs bosons discovered in 2012 in keeping with the predictions of the standard model of particle physics, or again the time for gravitational waves, first detected on 14 September 2015.

In order to become grown-up and gain their legitimacy, all great scientific discoveries have a vital need for mathematics, for algebraic equations and geometric figures. Mathematics has demonstrated its incredible power, and no serious physical theory today would dare to express itself in any other language.

CRYSTALLOGRAPHY

The mathematization of the world also impacts elegantly upon chemistry, where we shall now revisit some old acquaintances. At the start of the nineteenth century, the French mineralogist René Just Haüy, after accidentally dropping a piece of calcite, conjectured that the latter breaks into a myriad of splinters that all have the same geometric structure. The fragments were not random, and had plane faces, which formed quite precise angles with one another. For such a phenomenon to be produced, Haüy inferred that the piece of calcite must be formed from a multitude of similar elements that are assembled together in a perfectly regular manner. A solid with this property is called a crystal. In other words, a crystal observed at the microscopic scale consists of a pattern of several atoms or molecules that is identically repeated in all directions.

A pattern that is repeated? That sounds familiar. The principle bears an astonishing resemblance to the Mesopotamian friezes and the Arab tilings. A frieze has a pattern repeated in one particular direction, and a tiling has a pattern that is repeated in two directions. Thus, in order to study a crystal you need to go back to the same principles, but this time in three-dimensional space. The Mesopotamian artisans had discovered the seven categories of friezes, and the Arab artists the seventeen tilings. Using algebraic structures, it had become possible to prove that these numbers are indeed optimal: there are none missing. These same algebraic structures make it possible to establish that there exist 230 categories of tilings in 3D. The simplest ones include, for example, tilings with cubes, with hexagonal prisms, or with truncated octahedra,* as shown below.

From left to right: stackings of cubes, hexagonal prisms and truncated octahedra. These stackings can be extended to infinity in three-dimensional space.

In all cases, these figures stack and fit together perfectly without leaving holes, forming a structure that can be extended to infinity in all directions. Who would have thought that the geometrical thinking of the Mesopotamian artisans carried the seed for the basis of what would become an essential component of the study of the properties of matter?

---

* The octahedron is one of the five Platonic solids we met earlier. The truncated octahedron is obtained by cutting off the vertices of the octahedron in the same way in which we obtained the truncated icosahedron (or football) by cutting off the vertices of an icosahedron.

Crystals can be found almost everywhere in our daily life. Among other things, we might for example mention our table salt, which consists of a multitude of small crystals of sodium chloride, or quartz, whose very regular oscillations when an electric current is applied to it make it an indispensable element in our clocks and watches. However, some caution is advised, as the word 'crystal' is sometimes misused in everyday language; for example, crystal glasses are not in reality made of crystals in the scientific sense of the term.

If you would like to admire more spectacular specimens, you can always visit a mineralogy collection. That of the Université Pierre et Marie Curie in Paris is among the most beautiful in the world.

The dazzling effectiveness of the mathematization of the world, however, does not answer a disconcerting question. How is it that the language of mathematics is so perfectly appropriate for describing the world? In order to understand why this is so astonishing, let us return to Newton's formula

$$F = G \times \frac{m_1 \times m_2}{d^2}$$

The intensity of the gravitational force is given by a formula that involves two multiplications, a division and a squaring. The simplicity of this expression seems to be an unlikely stroke of luck. We are well aware, due to the number π and many others, that not all numbers can be expressed by simple mathematical formulae. From a statistical point of view, the complicated numbers are even much more numerous than the simple numbers. If you pick a number at random, you will have a much greater chance of coming up with a number with a decimal point than with an integer. Similarly, you will have a much greater chance of coming up with a number with an infinite rather than a finite decimal expansion, and a much greater chance of coming up with a number that is not expressible by any formula other than one that is calculable using only elementary operations.

Newton's formula is even more astonishing than that, because the force varies according to the masses of the objects and the distance between them. It is not a simple constant like π. However, whatever the masses of the two bodies are, and whatever the distance between them is, the attraction that they exert on each other is always measured by this same formula. Before Newton established his law, it would have been reasonable to suppose that there was absolutely no way of expressing the intensity of the force by a mathematical formula. And even if there had been such a way, one might well have expected a complicated formula involving much more monstrous operations than multiplications, divisions and squaring.

What a godsend it is that Newton's formula is what it is! And what a mystery it is that nature speaks the language of mathematics so fluently. It is often the case that models developed by mathematicians solely for their beauty find applications in the physical sciences centuries after they were developed. And this mystery does not end with gravitation. Electromagnetic phenomena, the quantum behaviour of elementary particles, and the relativistic spacetime deformation, are all examples of phenomena that can be expressed with a startling concision in the language of mathematics.

Take, for instance, the most famous of all formulae: $E = mc^2$. This equation, established by Albert Einstein, provides an equivalence between the mass and the energy of physical objects. We shall not explain the formula here – that is not our purpose. But just think of about it: this principle, which is generally considered to be one of the most fascinating and profound equations governing the operation of our universe, is expressed by an algebraic formula involving only five symbols. What is the reason for this miracle? Einstein is generally credited with the following phrase, which summarizes all that is stupefying about the situation: 'The most incomprehensible thing about the universe, is that it is comprehensible.' Here, comprehensible is taken to mean comprehensible by mathematics. In 1960, the physicist Eugene Wigner in turn spoke of the 'unreasonable efficiency of mathematics'.

So at the end of the day, do we know these abstract objects, numbers, figures, sequences and formulae so well that we think we have created them? If mathematics is really produced by our brain, why do we find it, like an errant spirit, outside our skulls? What is it doing in the physical world? Is it really there? Should we not perhaps see this ghost of reality as a sort of gigantic optical illusion? To imagine that mathematical objects have a form of existence outside the human mind would amount to giving them a reality although they are pure abstraction. What would the verb 'to exist' mean if we had to apply it to these objects that have nothing physical about them?

Don't expect me to come up with even the start of an answer to these questions!

14

THE INFINITELY SMALL

The close collaboration between mathematics and the physical sciences did not remain a one-way street for long. From the seventeenth century onwards, the two disciplines have never ceased to exchange ideas and feed off one another. Since physics is greedy for formulae, every new discovery now raises the question about the mathematics hidden behind it. Does it already exist or has it still to be invented? In the second scenario, mathematicians find themselves challenged to carve out bespoke new theories. In this way, they will discover the physical sciences to be one of their finest muses.

The development of Newtonian gravitation was one of the first discoveries to require innovative mathematics. In order to understand this, let us return to the trail of Halley's comet. It was one thing to know the force that attracts it towards the sun, but how could one use this knowledge to deduce its trajectory and useful information such as its position on a given date or its precise period of revolution?

In particular, one classical question that required an answer was that of determining the distance covered as a function of velocity. If I tell you that the comet moves in space with a velocity of 2,000 metres per second and I ask you the distance it will have covered in one minute, the answer is relatively simple. In one minute, the comet will cover 60 times 2,000 metres, that is to say 120,000 metres, or 120 kilometres. The problem is that the reality is more complicated than that. The comet's velocity is not fixed, but varies with time. At its aphelion, that is to say the point at which it is furthest from the sun, its velocity is 800 metres per second, whereas at its perihelion, when it is closest to the sun, its velocity is 50,000 metres per second. This is a huge difference!

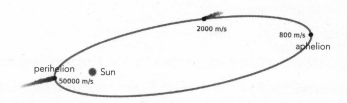

And all the subtlety stems from the fact that between these two extremes, the comet accelerates progressively, without retaining a fixed velocity at any moment in time. For example, there is a moment when the comet is moving at 2,000 metres per second, but that does not last. A fraction of a second beforehand, its velocity was a little greater, let's say 2,000.001, and a fraction of a second afterwards it has already fallen to 1,999.999. It is impossible to find even a shortest time interval, however tiny, during which the comet retains a constant speed. Under these conditions, how can we calculate the exact distance it covers?

To answer this question, mathematicians returned to a method strangely similar to that used by Archimedes 2,000 years earlier to calculate the number π. In the same way that Archimedes had approximated the circle by polygons having more and more sides, you can approximate the trajectory by considering that the comet moves with fixed levels of speed in increasingly shorter intervals. For example, you might imagine that the comet retains a fixed velocity of 800 metres per second for a certain time, then passes abruptly to 900 metres per second for a certain time, and so on. The trajectory calculated in this way will not be exact, but can be considered as an approximation. And, to increase the precision, all you need do is refine the speed levels. Instead of considering levels of 100 metres per second each, it is possible to descend in steps of 10, 1 or even 0.1 metres per second. The finer the steps in speed are, the closer the result to the actual trajectory of the comet.

The successive approximations obtained for the distance covered between aphelion and perihelion then form a sequence that might look as follows:

47   42   40   39   38.6   38.52   38.46   38.453 . . .

These numbers are given in astronomical units.* In other words, if you suppose that the velocity of the comet remains fixed in levels of 100 metres per second, you find that the distance between aphelion and perihelion is equal to 47 astronomical units. But this is still only a rough approximation. If you perform a refinement by taking levels of 10 metres per second, you find that this same distance is 42 astronomical units. As you refine the speed levels further and further, it

---

* The astronomical unit corresponds to the distance between the Earth and the Sun and measures approximately 150 million kilometres.

becomes clear that these lengths are growing closer and closer to a limiting value of around 38.45. This limiting value then corresponds to the actual distance covered by the comet between the two extreme points of its trajectory.

In some sense, we can venture to say that the limiting result corresponds to the result obtained by subdividing the comet's trajectory into infinitely many infinitely short intervals. In the same way, Archimedes' method for calculating π reduced to the assertion that a circle is a polygon with infinitely many infinitely small sides. The whole essence of these two assertions lies in the notion of infinity. We know since Zeno that infinity is an ambiguous and subversive notion and that our dealings with it can leave us teetering on the edge of paradox.

There are then two possible options: we can either categorically decline any involvement of infinity and find ourselves reduced to the laborious study of problems of Newtonian physics via limits of sequences of approximations; or we can take our courage in both hands and cautiously enter the mire of infinitely fine subdivisions. It was this second route that Newton opted to pursue in his *Principia mathematica*. He was followed by the German mathematician Gottfried Wilhelm Leibniz, who discovered the same concepts independently and developed more sharply certain notions that remained vague for Newton.

These explorations led to the birth of a new branch of mathematics now known as infinitesimal calculus.

The question of the paternity of infinitesimal calculus was debated at length in the years that followed. Although Newton was undoubtedly the first to set off down this road from 1669 onwards, it was a long time before he made his results public, and Leibniz pipped him to the post by publishing his work in 1684, three years before *Principia mathematica*. This tangling of dates could not help but trigger a keen debate between the Englishman and the German, who each claimed the invention of the theory for themselves, and even accused each other of plagiarism. However, it now appears as though neither of the two scholars was aware of the work of the other, and that they actually invented infinitesimal calculus independently.

As often happens at the birth of a theory, not everything was perfect from the start. There were numerous points in the works of Newton and Leibniz where rigour and justifications were lacking. Quite similarly to what had happened with imaginary numbers, certain methods were perceived to work while others did not, although there were no clear explanations as to why.

The object of infinitesimal calculus then became one of charting this still unknown territory by laying down permitted waypoints and others that, conversely, led to dead ends and paradoxes. In 1748 the Italian mathematician Maria Gaetana Agnesi published the *Instituzioni Analitiche*, or the *Analytical Institutions*, which provided a first full report on the state of the young discipline. A century later, it was the German Bernhard Riemann who completed the basic work on the subject, making the terrain navigable without danger.

From then, mathematicians moved en masse into infinitesimal calculus, and began to ask a whole lot of questions that went far beyond the original physical applications. For, far from being just a simple tool, the theory turned out to be exciting to analyse and marvellously beautiful. And since science is an endless game of ping-pong, these new developments gradually came to adorn new applications in domains other than astronomy.

Infinitesimals have since been harnessed in all problems that – like the trajectory of the comet – involve quantities that vary continuously. They are used in meteorology to model and predict the evolution of temperature or atmospheric pressure, in oceanography to track ocean currents, in aerodynamics to control how an aircraft wing or various space vehicles penetrate the air, and in geology to follow the evolution of the Earth's mantle and to study volcanoes, earthquakes or, over a longer term, tectonic drift.

During their explorations, mathematicians have discovered a multitude of strange results in the infinitesimal world, some of which have yielded utter bafflement.

One of the first ideas we may have when seeking to define an infinitely small interval is to take points. Euclid himself stated clearly that a point is the smallest geometrical element. With a length equal to 0, it is evidently infinitely small. Alas, this idea is too simple to work, and falls apart. To understand why, look at this line segment of unit length.

$$1$$

The segment consists of infinitely many points, each of which has length equal to 0. Thus, it appears possible to say that the length of the interval is equal to an infinity times 0. In algebraic language this is denoted by $\infty \times 0 = 1$, where ∞ is the symbol for infinity. The problem with this conclusion is that if we now consider an interval of length 2, that too consists of infinitely many points, which this time gives $\infty \times 0 = 2$. How can the same calculation have two different results? Letting the length of the interval vary, we can equally well obtain that $\infty \times 0$ equals 3, 1,000, or even π!

From this experiment we have to conclude that the concepts of zero and of infinity used in this context are not fine enough for the purpose that we wish to use them for. A calculation, such as $\infty \times 0$, whose result varies according to its interpretation, is called an indeterminate form. It is impossible to use these forms in algebraic calculations without finding that paradoxes show up at once in their thousands. If we were to authorize the multiplication $\infty \times 0$, then we would also have to accept that 1 is equal to 2, and other aberrations of a similar kind. In short, a different approach is required.

The second attempt might be as follows: since an infinitesimal interval cannot comprise a single point, it might be a segment delimited by two distinct but infinitely close points. The idea is attractive, but once again we hit a snag, because such points do not exist. The distance between two points can be as small as you like, but it will always retain a positive length. One centimetre, one millimetre, one billionth of a millimetre, or even less if you wish, are certainly all small lengths, but in no case are they infinitesimal. In other words, two distinct points never touch one another.

There is something very disconcerting about this statement. When you draw a continuous line, such as a segment, there are no holes in the segment and yet the points that make it up do not touch. No point is in direct contact with any other. The absence of holes in the line is solely due to the infinite accumulation of infinitely small points. And if we interpret the points of the line by their coordinates, the same phenomenon can be translated into algebraic terms: two different numbers never follow one another directly, there are always infinitely many numbers that slip in between them. Between the numbers 1 and 2 there is 1.5. Between the numbers 1 and 1.1, there is 1.05. And between the numbers 1 and 1.0001, there is 1.00005. We could keep it up like this for a long time. The number 1, like all the others, is not in contact with an immediate successor, and yet there is an endless accretion of numbers around it, thus ensuring a perfect continuity of their long succession.

After two fruitless attempts, we have to resign ourselves to admitting that these classical numbers as they had been defined up to that time are not powerful enough to engender infinitely small quantities. Thus, these elusive creatures that have a value of zero but are nevertheless smaller than all positive numbers will have to be created from scratch. This is what Leibniz and the scholars who followed in his footsteps did in constructing infinitesimal calculus. It took them three long centuries to define the rules of calculus that apply to these new quantities and to define the scope of these quantities' action. Thus, between the seventeenth and the twentieth century they produced a whole arsenal of theorems capable of answering the problems raised by infinitesimals with great efficiency.

The situation of numbers that are not really there, but that are used as intermediaries in calculations, starts to appear familiar. The negative and imaginary numbers came down this road. But, as on every occasion, the process of assimilation was long and it was difficult to predict the outcome. In the 1960s, the American Abraham Robinson initiated a new model, known as non-standard analysis, which incorporated the infinitesimals as numbers in their own right. However, unlike imaginary numbers, infinitesimal quantities, at the start of the twenty-first century, have still not really acquired the title of true numbers. Robinson's non-standard model remains marginal and is little used.

Perhaps there is still a need for discoveries, developments and remarkable theorems for the non-standard theory to become indispensable. Perhaps, on the other hand, it will never have the potential to become the dominant model, and the infinitesimals will then never be placed on an equal footing with their illustrious predecessors, the negative numbers and the imaginary numbers. Non-standard analysis is certainly beautiful, but perhaps not beautiful enough, and with too few benefits to arouse general enthusiasm. After just a few decades of existence, Robinson's model is still very young, and it is up to the mathematicians of the future to determine its fate.

Among the most fruitful developments in infinitesimal calculus, measure theory, conceived at the start of the twentieth century by the Frenchman Henri-Léon Lebesgue, is one of the most curious branches. The question raised is as follows: with the help of infinitesimals, is it possible to imagine and measure new geometric figures that have remained inaccessible to the ruler and compass? The answer is yes, and within just a few years these previously unknown figures sent even the most intuitive laws of classical geometry packing.

Let us take, for example, a segment graduated from 0 to 10.

$$0 \quad 1 \quad 2 \quad 3 \quad 4 \quad 5 \quad 6 \quad 7 \quad 8 \quad 9 \quad 10$$

0.1 φ π 7.28

Just as with Descartes, this graduation allows us to associate each point of the segment with a number between 0 and 10. On this segment, we can then distinguish the points that correspond to numbers that have infinitely many figures after the decimal point (such as π or the golden ratio φ). What then happens if we subdivide our segment according to this criterion? In other words, if we colour the points of the first category in a dark colour and the others in a light colour, what will the two geometric figures represented in dark and light colours look like?

There is no easy answer to this question, because these two categories of numbers are infinitely entangled. If you take an interval of numbers, however small, it will always contain both dark and light points. Between any two light points there is always at least one dark point, and between two dark points there is always at least one light point. Thus, the two figures look like lines of infinitely fine dust particles that fit perfectly into each other.

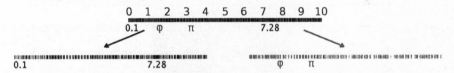

The segment [0,10] is subdivided into two parts: on the left are the numbers with a finite decimal expansion and on the right are those with an infinite decimal expansion.

The above representation is of course erroneous. It is only a rough visualization, since the details that are visible on it are drawn very small, but are not actually infinitesimals. In practice, it is impossible to draw these figures, which can only be properly apprehended through algebra and reasoning.

The following question then arises: how can one measure these figures? Since the initial segment has a length equal to 10, these two figures should conserve the same length between the two of them, but how is it shared out? Do they each have the same size of 5, or is one longer than the other? The answer that the mathematicians who studied this problem came up with is astonishing. Absolutely all the length is monopolized by the figure composed of numbers with an infinite expansion. The light figure has measure 10 and the dark figure 0. Although the two sets appear to be equal in their entanglement, there are infinitely many more light points than dark points.

Using Cartesian coordinates, powdery figures of this type can be generalized to surfaces and to volumes. We can, for example, consider the set of points of a square whose two coordinates have an infinite expansion.

Once again, this is a rough representation, which gives only a vague idea of the infinite precision of the details.

The measurement of powdery figures led to one of the most astounding results in mathematics. Despite all the best efforts of the mathematicians who studied this problem, some of these figures remain impossible to measure. This impossibility was demonstrated in 1924 by Stefan Banach and Alfred Tarski, who discovered a counter-example to the principle of the jigsaw puzzle.

They found a method for subdividing a ball into five fragments in such a way that these fragments could be reassembled to construct two balls rigorously identical to the first and with no holes.

The five intermediate figures they used are precisely powdery figures with infinitesimal subdivisions. If the pieces of the Banach–Tarski jigsaw puzzle were measurable, then the sum of their volumes would be equal both to the volume of the ball from which they originated and also to the volume of the two balls that they go on to form. Since that is impossible, there can be only one conclusion: the very notion of volume does not make sense for these figures.

In fact, the result of Banach and Tarski is much broader, since it asserts that if you take two classical geometrical figures in three dimensions, it is always possible to subdivide the first one into a certain number of powdery pieces, which can be used to reconstitute the second one. For example, a ball of the size of a pea can be subdivided into several fragments that can then be used to reconstitute a ball of the size of the Sun with no holes in the interior. This subdivision is often wrongly called the Banach–Tarski paradox because of its highly counter-intuitive nature. However, it is not a paradox, but a proper theorem made possible by the powdery figures with no contradictory reasoning.

Of course, the infinitesimal nature of these subdivisions makes them completely impossible to realize in practice. Powdery figures remain to this day in the cupboard of mathematical curiosities with no physical applications. Who knows whether they will be brought out of it one day to find unexpected uses?

15

MEASURING THE FUTURE

Marseille, 8 June 2012.

This morning, I got up at first light. A little nervy, but burning with impatience, I wolfed down my breakfast, pulled on my best shirt,* and then I was off. Outside, the sun's first rays were lighting the skies above Provence and the coolness of the night was vanishing rapidly. It promised to be a hot day. The fish market was setting up on the Vieux-Port, while a few early-morning tourists were already strolling along La Canebière.

There was no time for slacking today. I went down into the Métro and headed quickly off towards the Château-Gombert neighbourhood, in the north of the city. That's where the Centre for Mathematics and Information Technology (CMI), where I have now been working for four years, is located. Each day, a hundred-odd of mathematicians work here. On arriving in my office I checked my materials one last

---

* My only one, if the truth be known.

time: three wide hemispherical receptacles full of multicoloured balls, and beside them a stack of photocopied notes.

Today is my last day at the CMI. This afternoon at 2 o'clock, I will have my viva, where I defend my doctoral thesis in mathematics on interacting urns.

The thesis years form an atypical period in the life of a scientist. While they are still students on paper, those preparing a thesis no longer have courses to follow or any termly examinations to sit. In reality, our days bear a much greater resemblance to those of fully-fledged researchers. Thus, a typical schedule might include reading the latest articles, discussions with other mathematicians, taking part in seminars, and working to make progress in one's field, to come up with conjectures, to fashion new theorems, to prove them, and to publish them. All this takes place under the supervision of a seasoned mathematician who is responsible for guiding our first steps in the world of research and teaching us the tricks of the trade. In my case, my thesis director has been the Franco-Croatian mathematician Vlada Limic, a specialist in the subject area in which I have carried out my research over these past four years. Her work and mine fall within a branch of mathematics that was born in the middle of the seventeenth century: probability.

In order to understand what this discipline is about, we again have to delve back into the depths of history. So while I have to wait until 2 p.m., let us leave the CMI for a while and let me lead you along the adventure-filled paths of stochastics.

I have been fascinated by chance for a long time. Already in prehistoric times human beings observed a multitude of unexplained, irregular phenomena with no apparent causes, which nature faced them with. At first, and for lack of anything better, they accused the gods. Eclipses, rainbows, earthquakes, epidemics, exceptional floods on rivers, and comets were manifestations that were interpreted as divine messages addressed to whosoever could decipher them. That task was entrusted to sorcerers, oracles, priests or other shamans who, as they had to earn their living, developed while they were at it a whole panoply of rituals designed to consult the gods without waiting for them to deign to manifest themselves. In other words, humans began to imagine means of creating random phenomena on demand.

Belomancy, or the art of divination by the use of arrows, is one of the most ancient witnesses to this. Inscribe on arrows the various possible answers to Frequently Asked Questions that you are addressing to your god; place the arrows in your quiver; shake the whole thing and draw out an arrow at random: there's their reply. It was in this way, for example, that Nebuchadnezzar II, king of Babylon, chose the enemies upon whom he declared war in the sixth century BC. Apart from arrows, the objects drawn could take many forms: stones, tablets, wands or coloured balls. The Romans used the word *sors* to refer to these objects. The English expression 'by sort', meaning by the casting or drawing of lots, comes from this, as does the word 'sortilege', which originally referred to one who practises divination or sorcery.

Gradually, the mechanisms for performing random draws proliferated and found numerous applications. Several political systems used them, as in Athens to designate the five hundred citizens sitting in the boule, the council of citizens, or several centuries later in Venice, in the process of appointing the Doge. Chance has also been a great source of inspiration for the creators of games. Think, for example, of the invention of 'heads or tails', of the numbered dice to which the Platonic solids have lent their shapes, or again of card games.

It was in fact through games of chance that the decisions of the gods finally attracted the attention of certain mathematicians. These people had the strange idea of playing at being the measurers of destiny by studying the properties of the future before it arrives, using logic and calculus.

It all began in the mid-seventeenth century, at a meeting of the Académie parisienne, a forerunner of the French Academy of Sciences, created in 1635 by the mathematician and philosopher Marin Mersenne. During a discussion between scholars from a variety of disciplines, the writer Antoine Gombaud, who spent his spare time with mathematics, submitted to the assembly a problem that had occurred to him. Imagine, he said, that two players have staked a certain amount of money in a gambling game which has a series of legs in which the winner is the first to achieve three wins, but that the game is interrupted when the first player is leading by two legs to one. How should these two players share out the stakes before leaving?

Among the scientists present on that day, the problem attracted the particular attention of two Frenchmen: Pierre de Fermat and Blaise Pascal. After a few exchanges of letters, they both came to the conclusion that three-quarters of the stake should go to the first player and the remaining quarter to the second.

To reach this answer, the two scholars listed all the scenarios that could have occurred in the game had it run to its end, and worked out the chances of each of them occurring. Thus, in the next hypothetical leg, the first player would have had a 50 per cent chance of winning the game, while the second player would have had a 50 per cent chance of returning to equality. And in this second eventuality, a new leg would then have been played, with each player having an equal chance of winning, which then gives two scenarios each with a 25 per cent chance of occurrence. This reasoning can be translated into the following graph, which summarizes the various possible outcomes for the game:

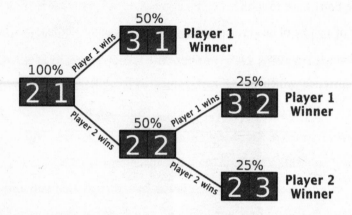

In short, we see that 75 per cent of the outcomes lead to victory for the first player while only 25 per cent lead to victory for the second. Pascal and Fermat's conclusion thus shares the stakes in these same proportions: it is proper that the first player should retain 75 per cent of the stakes and the second the remaining 25 per cent.

The reasoning of the two French scholars turned out to be particularly fruitful. Most games of chance can be subjected to an examination of this type. The Swiss mathematician Jacob Bernoulli was one of the first to follow in their footsteps, when at the end of the seventeenth century he wrote a work entitled *Ars conjectandi*, or *The Art of Conjecturing*, which was only published posthumously in 1713. In that book, he returned to the analysis of classical games of chance and stated for the first time one of the fundamental principles of probability theory: the law of large numbers.

This law asserts that when a random experiment is repeated a large number of times, the mean value of the results becomes increasingly predictable and tends to a limiting value. In other words, even the most random random process, in the long term, gives rise to average behaviours that no longer have anything random about them.

We don't have to look very far to grasp this phenomenon. A simple study of a game of heads or tails shows the emergence of the law of large numbers. If you toss a balanced coin, each side has a 50 per cent chance of occurrence, which can be represented by the following histogram.

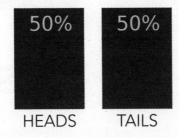

Imagine now that you toss a coin twice in succession and that you count the total number of heads and tails. There are then three possibilities: either two heads, or two tails, or a head and a tail. It might be tempting to think that these three eventualities occurred in equal proportions, but that is not the case. In reality, there is a 50 per cent chance of obtaining a head and a tail, while the probabilities of two heads or two tails are each 25 per cent.

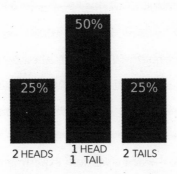

This imbalance is due to the fact that two different draws can give the same final result. When you toss the coin twice, there are in fact four possible scenarios: head–head, head–tail, tail–head and tail–tail. The scenarios head–tail and tail–head give the same final result of one head and one tail, which explains why this eventuality is twice as likely. Similarly, players are well aware that if you throw two dice, their sum has a greater chance of being equal to 7 than to 12, since there are several ways of obtaining 7 (1 + 6; 2 + 5; 3 + 4; 4 + 3; 5 + 2 and 6 + 1), while there is only one way of obtaining 12 (6 + 6).

The more one increases the number of tosses, the more accentuated this phenomenon becomes. The scenarios that differ from the average gradually become an ultra-minority in comparison with the average scenarios. If you toss a coin ten times in succession, there is an approximately 66 per cent chance that you will toss between four and six heads. If you toss this same coin one hundred times, you will have a 96 per cent chance of obtaining between forty and sixty heads. And if you toss it one thousand times you will have a 99.99999998 per cent chance of tossing between 400 and 600 heads.

When you draw the histograms corresponding to 10, 100 and 1,000 tosses, it emerges that, little by little, the great majority of possible outcomes close in on the central axis, to the point where the rectangles corresponding to extreme situations become invisible to the naked eye.

In short, the law of large numbers asserts that when a random experiment is repeated indefinitely, the average value of the results obtained inevitably converges to a limiting value that has nothing random about it.

This principle underlies the operation of polls and other statistics. In a population, take 1,000 people and ask them whether they prefer

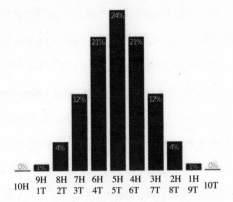

Probability histogram for the possible scenarios when 10 coins are tossed.

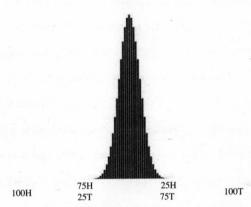

Probability histogram for the possible scenarios when 100 coins are tossed.

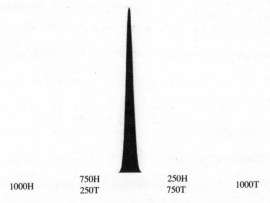

Probability histogram for the possible scenarios when 1,000 coins are tossed.

dark chocolate or milk chocolate. If 600 reply that they prefer dark and 400 prefer milk chocolate, there is every chance that in the population as a whole, even if it comprises millions of individuals, the proportion will also be close to 60 per cent preferring dark and 40 per cent milk chocolate. Asking one person selected at random about his or her tastes can be considered to be a random experiment in the same way as tossing a coin. The options of heads or tails are simply replaced by dark or milk chocolate.

Of course, it would have been possible to have been unlucky and to have happened upon 1,000 people who all liked dark chocolate or 1,000 people who all liked milk chocolate. But these extreme scenarios all have an absolutely tiny chance of occurrence, and the law of large numbers guarantees that if we question a large enough sample of the population, the average value obtained will have a very strong chance of being close to the average value for the population as a whole.

Pushing the decoding of multiple scenarios and their chances of occurrence further, it is also possible to establish a confidence interval and to estimate the risks of error. One might say, for example, that there is a 95 per cent chance that the proportion of the population preferring dark chocolate lies between 57 per cent and 63 per cent. Any properly conducted poll should then always be accompanied by such figures that indicate its accuracy and its reliability.

PASCAL'S TRIANGLE

In 1654, Blaise Pascal published a work entitled *Traité du triangle arithmétique* or *Treatise on the Arithmetical Triangle*. In it he described a triangle consisting of cells with numbers inscribed inside them.

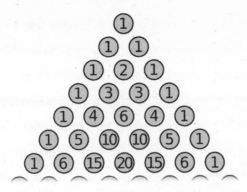

Only the first seven lines are shown here, but the triangle can be extended to infinity. The numbers in the cells are determined by two rules. First, the cells on the edges contain only 1. Second, the cells in the interior contain the sum of the two cells immediately above them. For example, the number 6 on the fifth line is equal to the sum of the two 3s that lie above it.

In truth, this triangle was already known well before Pascal became interested in it. The Persian mathematicians al-Karaji and Omar Khayyam referred to it in the eleventh century. In the same period, it was studied in China by Jia Xian, whose works were extended in the thirteenth century by Yang Hui. In Europe, Tartaglia and Viète also knew about it. However, Blaise Pascal was the first to devote such a detailed and complete treatise to it. He was also the first to discover the existence of a close connection between the triangle and the enumeration of possible outcomes in probability.

Each line of Pascal's triangle can in fact be used to count the number of possible scenarios in a succession of events with two outcomes such as heads or tails. If you toss a coin three times in succession, then there are eight possible scenarios: head–head–head, head–head–tail, head-tail-head, head-tail-tail, tail-head-head, tail-head-tail,

tail–tail–head and tail–tail–tail. When you do the sums, you realize that of these eight possible scenarios:

- **1 SCENARIO GIVES THREE HEADS;**
- **3 SCENARIOS GIVE TWO HEADS AND ONE TAIL;**
- **3 SCENARIOS GIVE ONE HEAD AND TWO TAILS;**
- **1 SCENARIO GIVES THREE TAILS.**

Now this sequence of numbers, 1–3–3–1, corresponds exactly to the fourth line of the triangle. This is not random, and Pascal succeeded in proving it.

Looking at the sixth line, you can, for example, see that when a coin is tossed five times, there are ten scenarios that give 2 heads and 3 tails. Going further into the triangle, it becomes easy to count the scenarios resulting from ten tosses of a coin: they are given on the eleventh line. One hundred tosses are given by the 101st line, and so on. It is also thanks to Pascal's triangle that the histograms presented earlier were easy to draw. Without that, the number of outcomes becomes so prodigiously large that it fast becomes impossible to list them all individually.

Outside probability, Pascal's triangle has also revealed numerous connections with other domains of mathematics. The numbers in it are, for example, of great use in the algebraic manipulations that make it possible to solve certain equations. In its boxes we can also find several well-known sequences of numbers such as the triangular numbers (1, 3, 6, 10 . . .), which lie on one of its diagonals, and the Fibonacci sequence (1, 1, 2, 3, 5, 8 . . .), which is obtained by adding the terms along parallel inclined lines.

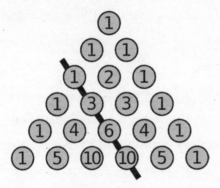

The sequence of triangular numbers in Pascal's triangle

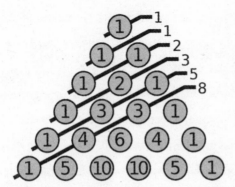

The Fibonacci sequence in Pascal's triangle

In the centuries that followed, probability theory developed increasingly refined and powerful tools with which to analyse the set of all possible outcomes. Soon, a close and fruitful collaboration was established with infinitesimal calculus. In fact, numerous random phenomena produce outcomes that may be subject to infinitely small variations. In a meteorological model, for example, the temperature varies continuously. Just as a segment has a length while the points that comprise it do not, certain events may occur although each of the outcomes that comprise them has no chance of occurring individually. The probability that, in a particular week, the temperature is exactly 23.41

degrees (or any other precise value) is equal to zero. However, the global probability that the temperature lies between 0° and 40° is definitely positive.

Another challenge for probability theory was to understand the behaviour of random systems that are capable of modifying themselves. A coin remains the same, whether it has been tossed once or a thousand times, but many real-life situations are not so simple. In 1930, the Hungarian mathematician George Pólya published an article in which he sought to understand the propagation of an epidemic within a population. The subtlety of this model stems from the fact that an epidemic propagates faster when a large number of people are affected by it.

If a large number of people around you have an infection, you will have a greater chance of becoming infected yourself in turn. And if you become infected, it is you who will increase the risks for the people around you. In short, the process is self-reinforcing and the probabilities are in permanent evolution. The process of infection is said to be reinforced.

Subsequently, numerous variants of reinforced random processes were developed, and there were multiple applications. One of the most fertile of these was their use in population dynamics. Take an animal population whose biological or genetic characteristics you wish to follow across the generations. Imagine, for example, that 60 per cent of its individuals have brown eyes and 40 per cent blue eyes. Then, by heredity, the new individuals who are born have a 60 per cent chance of having brown eyes and a 40 per cent chance of having blue eyes. The evolution of eye colour in this population thus has a dynamic similar to the propagation of an epidemic: the more common a certain colour is, the greater the chances that this colour will appear again, and thereby increase its proportion. The process is self-reinforcing.

Thus, the study of Pólya's model makes it possible to gauge the evolutionary probabilities of various biological characteristics of species. Some characteristics may eventually disappear. Others, conversely, may come to dominate in the population as a whole. Others again may reach an intermediate equilibrium in which their proportion is subject only to small variations over the generations. It is not possible to know in advance which of these scenarios will arise, but, just as in the game of heads or tails, the probabilities make it possible to identify the dominant future outcomes and to predict the most likely evolution in the long term.

When George Pólya died in 1985, I was scarcely one year old. So I can say that I was a contemporary, for a few months, of the man who initiated the theory on which I would work myself and in which I would discover several theorems.

Without going too deeply into the details, my results concern the evolution of several reinforced random processes that interact occasionally. Imagine, for example, several groups of animals of the same species living separately in the same territory, but now and then admitting the migration of a few individuals from one group to another. What are the possible future outcomes and how can we calculate their probabilities? These are questions to which my research has provided some partial answers.

Oh, of course, my theorems are modest, and it is audacious to dare to mention them in this epic saga, which comprises so many great names. While I believe I have been a proper researcher going about my work correctly in the course of my four-year thesis, my discoveries pale in comparison with those of other much more brilliant mathematicians than myself. They were however sufficient to convince the jury to which I spent an hour explaining them on that 8 June 2012 to award me the title of doctor.

It is quite moving to enter, through this ceremony, into a stream of history so distinguished. The word 'doctor' comes from the Latin *docere*, meaning 'to teach'. A doctor is thus someone who has acquired a sufficient mastery of their subject to be able to pass it on in their turn. Since the end of the Middle Ages, the universities, the modern heirs of the Mouseion of Alexandria or the Bayt al-Hikma in Baghdad, have delivered doctorates and provided a stable and permanent institutional framework for scientific research and teaching.

Since then the sciences have initiated a movement which, down the centuries, has seen researchers, teachers and pupils succeed one another in an almost constant unfolding of the generations. It is a source of amusement in all this that we can trace the academic descent of scientists. While my thesis director was the mathematician Vlada Limic, she herself had the British probabilist David Aldous as her thesis director several years earlier. And we could go a long way down this path. By stepping back from pupil to master, you can thus retrace the complete 'genealogy' of mathematicians. Look here at my line – it goes back to the sixteenth century over more than twenty generations!

My most distant ancestor is thus the mathematician Niccolò Tartaglia, whom we have already met. It is impossible to go back further because the Italian scholar was self-taught. He came from a poor family, and legend even has it that the young Tartaglia had to steal the books from which he learnt mathematics from his school.

In this genealogy, you will find Galileo and Newton, who need no further introduction. In a corner, you will also see Marin Mersenne, who created the Académie parisienne where probability theory was born. His pupil Gilles Personne de Roberval was the inventor of the double-beam balance that carries his name. A little further on, George

Darwin was the son of Charles Darwin, the father of the theory of evolution.

There is nothing particularly exceptional about finding such celebrities in this line – most mathematicians whose genealogy goes so far back are able to find great names in it. Moreover, it has to be said that this diagram shows only my direct ancestors and ignores my very numerous 'cousins'. Today, Tartaglia has more than thirteen thousand descendants, and this number continues to grow with each year.

16

THE COMING OF MACHINES

The Paris Métro's Arts et Métiers station is among the city's strangest. For travellers descending into it, it is like being swallowed up in the copper innards of a gigantic submarine. Large reddish gearwheels hang from the ceiling and the platform walls are lined with a dozen portholes. Look through them and you discover curious depictions of various old or unusual inventions. You can find elliptical gear wheels, a spherical astrolabe and waterwheels alongside a dirigible airship or a steel converter. In the constant stream of busy Parisians who rush and surge through the underground corridors, you would scarcely be surprised to see the imposing figure of Jules Verne's Captain Nemo appear right out of the book.

The décor of the Métro is, however, only a taste of what awaits on the surface. Today I'm heading for the Conservatoire national des arts et métiers (French national conservatory of arts and crafts), whose museum houses one of the most important collections of old machines of all kinds. From the first motor cars to dial telegraphs, via piston manometers, Dutch automaton clocks, voltaic pile batteries, Jacquard looms, screw printing presses and siphon barometers, all these inventions rescued from the past carry me into the astonishing technological maelstrom of the last four centuries. Hanging over the middle of the grand staircase there is a nineteenth-century aeroplane that looks like a gigantic bat. Taking a detour down a corridor, I come face to face with Lama, the first robot designed by Russian scientists in the twentieth century to rove the surface of Mars.

I quickly pass by all these fabulous objects and make my way directly to the second floor, which houses the gallery of scientific instruments. This gallery contains telescopes, clepsydra, compasses, Roberval balance scales, gigantic thermometers and sublime astronomical globes pivoting on their axes. Then suddenly, in the corner of a display cabinet, I spot what I have come to see: the Pascaline. This curious machine looks like a brass jewel box, measuring 40 by 20 centimetres, with six numbered wheels fixed to its upper surface. The mechanism was designed in 1642 by Blaise Pascal, who was then aged only nineteen. In front of me lies the very first calculating machine in history.

But was it really the first? In truth, devices that could be used to perform calculations were in existence well before the seventeenth century. One might say that human fingers formed the first calculating machine of all time, and that *Homo sapiens* used various accessories for counting in very early times. The Ishango bones and their notches, the clay counters of Uruk, the wands of the ancient Chinese, and also the abacuses that were very successful from antiquity onwards, were all instruments used to support counting and calculation.

However, none of them fits the conventional definition of a calculating machine.

To understand the operation of the Pascaline, let us take a few moments to describe how a classical abacus works. The object consists of several rods with beads running on them. The first rod corresponds to units, the second to tens, the third to hundreds, and so on. Thus, if you want to represent the number 23, you push two beads along the tens column and three along the units column. And if you want to add 45, you push four additional tens and five additional units, which gives 68.

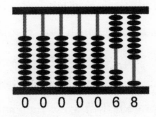

0 0 0 0 6 8

If, on the other hand, the addition involves a carry, then a small extra procedure is required. If you now want to add 5 to 68, you only have one remaining ball on the units rod. In this case, once you've reached the unit number 9, you need to pull back down all the beads from that rod to start over from scratch (0), all the while placing a bead in the carry that has the tens column. You have now reached the answer, 73.

This procedure is not very complicated. However, it is this that prevents us from referring to the abacus, and to all the mechanisms preceding the Pascaline, as calculating machines. To carry out the same operation, the user does not perform the same actions regardless of whether or not there is a carry. In reality, the machine is just an aide-memoire, which reminds its human operator where they have got to, but always leaves them to perform the various stages of the calculation manually. When, on the other hand, you perform an addition on a modern calculator, you don't have to worry at all about how the machine finds the result. There may or may not be carries, but it's not your concern. It is no longer necessary to think or adapt to the situation, the device looks after everything.

According to this criterion, the Pascaline therefore actually is the first calculating machine in history. Although the mechanism is very precise and required great skill on behalf of its constructor, its operational principle remains very simple. On the upper face of the machine there are six wheels with numbered notches.

The first wheel on the right represents the units figure, the second the tens figure, and so on. Above the wheels is the display area, consisting of six small boxes, one for each wheel, each of which indicates a figure. To insert the number 28, simply turn the tens wheel clockwise by two notches and the units wheel by eight notches. Based on an internal system of gears, the figures 2 and 8 will then be displayed for you in the two corresponding boxes. And now, if you wish to add 5 to this number, there's no need to perform the carry yourself: simply turn the units wheel by five notches, and when this passes from 9 to 0, the tens figure will automatically go from 2 to 3. The machine will now display 33.

And this works with as many carries as you wish. Display 99,999 on the Pascaline, then turn the units wheel by one notch. You will see all the carries cascade to the left, leading to the appearance of the number 100,000, without the user having to do anything.

After Pascal, other inventors perfected his machine to enable it to perform more and more operations increasingly rapidly and efficiently. At the end of the seventeenth century, Leibniz was one of the first to follow in his footsteps by designing a mechanism that made it possible to do multiplication and division more easily. However, his system was never completed, and the machines he produced still made carry errors in several special cases. His ideas were not fully carried out until the eighteenth century, when multiple prototypes, each one more reliable and more efficient, were created by even more ingenious and imaginative inventors. However, the increasing complexity of the mechanisms was at the expense of the size of the machines, which from being objects of modest dimensions, soon became small pieces of furniture.

In the nineteenth century, calculating machines moved out of their niche and became widespread in a similar manner to their cousins, the typewriters. Many accountancy firms, businessmen or simply traders, acquired these calculators, which became part of the furniture and rapidly made themselves indispensable. It makes you wonder how people had managed until then.

Continuing my visit to the museum, I come across several successors to the Pascaline, such as Thomas de Colmar's arithmometer, Léon Bollée's multiplying machine, Dubois's polychromatic arithmograph, and Felt and Tarrant's comptometer. One of the most successful mechanisms was the arithmometer developed in Russia by the Swedish engineer Willgodt Theophil Odhner. This machine consists of three main elements: the top part, where you pull small levers to indicate the number you wish to operate on; the bottom part, which consists of a carriage that can be shifted horizontally, on which the result of the operation is displayed; and the handle, which is used to perform the operation.

With each turn of the handle, the number indicated in the top part is added to the number already displayed on the carriage at the bottom. To perform a subtraction, simply turn the handle in the other direction.

Now imagine that you wanted to perform the multiplication of 374 × 523. Enter the number 374 in the top part and turn the handle three times. The bottom part then displays 1,122, the result of the operation 374 × 3. Now shift the display carriage by one notch in the tens direction and turn the handle twice. The number 8,602, which corresponds to 374 times 23, is displayed. Shift the carriage again by one notch to reach the hundreds, turn the handle five more times, and there is your result: 195,602. After a little practice and training, you will only need a few seconds to perform your multiplication.

In 1834, the British mathematician Charles Babbage had an idea that was to say the least whimsical, namely to cross a calculating machine with a loom. Over a span of a few years, several improvements to the operation of looms had been made. One of these was the introduction of punched cards that made it possible for a single machine to produce a great variety of patterns without the need to change its settings. Depending on whether or not there is a hole at a particular place in the card, an articulated hook attached to a harness carrying the warp thread is raised or lowered, and the weft thread passes above or below the warp thread. In short, you have only to record the desired pattern on the punched card and the machine then adjusts itself accordingly.

Based on this model, Babbage devised a mechanical calculator that would not be applied to performing certain fixed calculations, such as addition or multiplication, but would be capable of adapting its behaviour and of performing millions of different operations as a function of a punched card that the user would insert in it. More precisely, this machine was able to perform all polynomial operations, that is to say calculations that combine the four basic operations and the taking of powers in an arbitrary order. As with the way that the Pascaline enabled its user to make the same movements whatever numbers were used, Babbage's machine enabled you to make the same movements whatever operations were to be carried out. There was no longer a need, as had been the case, for example, for Odhner's calculator, to turn the handle in one direction or the other, depending on whether you were adding or subtracting. You just had to enter your calculation on the punched card and the machine would take care of everything. This revolutionary form of operation made Babbage's machine the very first computer in history.

Nevertheless, its operation raised a new challenge. In order to perform a calculation, you had to equip the machine with the proper card. This card consisted of a succession of holes or non-holes, that the machine would detect and that would inform it, stage by stage, which operations should be performed. Thus, before starting, the user of the machine had to translate the calculation that they wished to perform into a succession of holes or non-holes to be 'read' by the machine.

This translation work was pursued and developed by the British mathematician Ada Lovelace. She studied the operation of the machine in depth and understood its full potential, possibly more than Babbage himself had ever imagined. In particular, she described a complicated code capable of calculating the Bernoulli sequence (which was discovered a century earlier by the Swiss Jacob Bernoulli and is extremely useful in infinitesimal calculus). This code is generally considered to be the very first computer program, and makes Lovelace the first programmer in history.

Ada Lovelace died in 1852 at the age of thirty-six. Charles Babbage tried all his life to construct his machine, but died in 1871 before his prototype was completed. The first working Babbage machine was not seen until the twentieth century. There is something both impressive and magical about observing one of these calculators in operation. Its large size (around 2 metres high and 3 metres wide) and the choreographed ballet of the hundreds of gearwheels that whirr and whirl inside it are dazzling and awe-inspiring.

The British scholar's unfinished prototype is now housed in the Science Museum in London, where it can still be admired. Demonstrations of a working example that was assembled at the start of the twenty-first century can be seen at the Mountain View Computer History Museum in California.

The twentieth century saw the triumph of computers on a scale that Babbage and Lovelace would surely never have imagined. Calculating machines benefited from convergent spin-offs from both the oldest and the most recent mathematics.

On the one hand, infinitesimal calculus and imaginary numbers provided a mathematical approach to electromagnetic phenomena, which soon led to the birth of electronic devices. On the other hand, the nineteenth century saw a rebirth of interest in questions relating to the foundations of mathematics, to axioms and to elementary reasoning that can be used to produce proofs. The first case provided machines with a physical infrastructure with an extraordinary speed, while the second provided for an efficient organization of elementary calculations in order to produce the most complicated results.

One of the main architects of this revolution was the British mathematician Alan Turing. In 1936 Turing published an article in which he established a parallel between the provability of a theorem in mathematics and the computability of a result by a machine in computer science. In that paper he was the first to describe the operation of an abstract machine, to which he gave his name and which is still widely used in information theory. The Turing machine is purely imaginary. The British mathematician was not concerned with the concrete mechanisms by which it could be constructed. He simply stated the elementary operations that his machine was able to carry out, then asked what it was capable of obtaining by combining these. Here, one can clearly see the analogy with a mathematician stating his or her axioms and then attempting to deduce theorems from them by combining them.

The sequence of instructions that one gives to a machine in order to end up with a result is called an algorithm – the word is a Latin adaptation of the name 'al-Khwārizmī'. It has to be said that computer algorithms were in the first instance largely inspired by procedures for solving problems that were already known to the ancients. You will remember that in his *al-jabr* al-Khwārizmī not only considered abstract mathematical objects, but also gave practical methods that enabled the citizens of Baghdad to find the solutions to their problems without needing to understand the whole theory. Likewise, a computer does not need to have the theory explained to it, and in any case cannot understand it. It simply needs you to tell it which calculations need to be performed and in what order.

Here is an example of an algorithm that can be input to a machine. The latter has three memory cells in which numbers can be written. Can you guess what this algorithm will calculate?

- **STEP A.** WRITE THE NUMBER 1 TO MEMORY CELL 1, THEN GO TO STEP B.
- **STEP B.** WRITE THE NUMBER 1 TO MEMORY CELL 2, THEN GO TO STEP C.
- **STEP C.** WRITE THE SUM OF MEMORY CELL 1 AND MEMORY CELL 2 TO MEMORY CELL 3, THEN GO TO STEP D.
- **STEP D.** WRITE THE NUMBER IN MEMORY CELL 2 TO MEMORY CELL 1, THEN GO TO STEP E.
- **STEP E.** WRITE THE NUMBER IN MEMORY CELL 3 TO MEMORY CELL 2, THEN GO TO STEP C.

You will note that the machine will go round in a loop, because Step E returns to Step C. Steps C, D and E will then be repeated *ad infinitum*.

So where is this leading? What does this machine do? It takes a little thought to decrypt this sequence of instructions, which is given cold and without explanations. However, you should be able to see that this algorithm calculates numbers with which we are already familiar, namely the terms of the Fibonacci sequence.*

Steps A and B initialize the first two terms of the sequence: 1 and 1. Step C calculates the sum of the two previous terms. Steps D and E then move the results obtained within the memory so that the process can recommence. If you observe the data items that are successively displayed in the memory cells as the machine operates, you will then see the following numbers stream by: 1, 1, 2, 3, 5, 8, 13, 21, and so on.

While this algorithm is relatively simple, it is still not simple enough to be read by a Turing machine. As defined by their creator, these machines are actually not capable of performing an addition, as is the case in Step C. They are only able to read from the set position in the memory to which they currently point, to write to this set position or to change this set position. However, they can be taught to add by inputting an algorithm in which figures are added digit by digit using carries, as in the abacus. In other words, addition is not an axiom of the machines, but is already one of the theorems for which an algorithm must be given before one can use it. Once this algorithm has been written, it has only to be substituted for Step C to enable a Turing machine to calculate the Fibonacci numbers.

With increasing complexity, the Turing machine can then be taught to perform multiplications and divisions, to take squares and square roots, to solve equations, to calculate approximations for π or trigonometric ratios, to determine the Cartesian coordinates of geometric figures, or even to carry out infinitesimal calculus. In short, provided

---

* Recall that the first two terms of the Fibonacci sequence are 1 and 1, then each term is the sum of the two previous ones. The sequence begins as follows: 1, 1, 2, 3, 5, 8, 13, 21 . . .

it is given the appropriate algorithms, a Turing machine can do all the mathematics we have explored up to now with far greater accuracy.

THE FOUR-COLOUR THEOREM

Take the map of a territory consisting of several regions delimited by their regional boundaries. What is the minimum number of colours needed in order to be able to colour this map so that two adjacent regions never have the same colour?

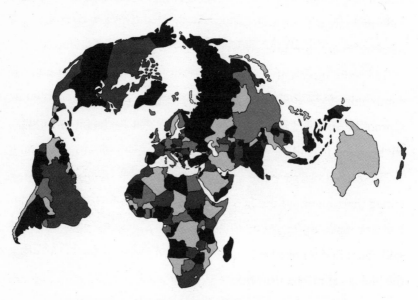

In 1852, the South African mathematician Francis Guthrie investigated this question and conjectured that whatever the map is, it is always possible to use just four colours at the most. After him, many scholars attempted to prove this statement, but no one succeeded in doing so for more than a century. Some progress was made, however, and it was established that all the possible maps could be reduced to 1,478 particular instances, where each of these required numerous verifications. But the point was that it was impossible for a human being, and even for a whole team of human beings, to carry out all these verifications themselves. A whole lifetime would not have been enough. Imagine if you can the frustration of these mathematicians, who had within their grasp the method that would enable them to prove or refute the conjecture, but were unable to use it because of time constraints.

In the 1960s a few researchers began to think of involving a computer, and in 1976 it was two Americans, Kenneth Appel and Wolfgang Haken, who finally announced that the theorem had been proved. Even so, it took more than 1,200 hours of computation and 10 billion elementary machine operations to get through the 1,478 maps.

The announcement came like a bombshell to the mathematical community. How should this new form of 'proof' be received? Should a proof be accepted as valid if it is so long that no one human being has been able to read it all? How far can one have confidence in machines?

These questions gave rise to many debates. While some people maintained that you could never be 100 per cent certain that the machine had not made mistakes, others replied that the same thing could be said about human beings. Is an electronic mechanism less worthy than the biological mechanism that is a *Homo sapiens*? Is a proof produced by a metallic machine less reliable than a proof provided by an organic machine? Mathematicians, and sometimes the greatest among them, have often turned out to have made mistakes that were not detected until much later on. Should that make us question whether the whole edifice of mathematics is well founded? There is no doubt that a machine may have bugs and may sometimes make mistakes, but while its reliability is at least equal to that of a human being (and it is sometimes greater), there is no reason to reject its results.

Today, mathematicians have learnt to have confidence in computers, and most now accept the validity of the proof of the four-colour theorem. Numerous other results have since been proved using computers. However, this type of method is not always well appreciated. A concise proof hand-crafted by a person is often considered to be more elegant. While the aim of mathematics is to understand the abstract objects that it manipulates, man-made proofs are often more instructive and generally provide a better insight into the deep meaning.

On 10 March 2016, the world had its eyes fixed on Seoul. It was there that the much-awaited match of the game of Go between the best player in the world, the South Korean Lee Sedol, and the AlphaGo computer was being held. The match, which was streamed live on the Internet and broadcast on several television channels, was followed by hundreds of millions across the world. The atmosphere was tense. A computer had never beaten a human being at this level.

Go is known to be one of the most difficult games for a machine to learn. Its strategy requires the players to be highly intuitive and creative. But, while machines are very good at calculations, it is very much harder to find algorithms that simulate instinctive behaviours. Other famous games, such as chess, are more calculation-based. That is why the Deep Blue computer managed to beat the Russian chess grandmaster Garry Kasparov in 1997 in a match that also caused a great stir. In other games such as draughts, computers have even perfected an unbeatable strategy. No human being can hope to beat a computer at draughts. A drawn match might just be possible if the human being plays to perfection. Within the family of great games of strategy, in 2016 Go still remained the last to resist the assault by machines.

At an hour into the match, we had reached the thirty-seventh move and the match seemed tight. It was then that AlphaGo astonished all the specialists who were following the match. The computer decided to play its black stone in position O10. On the Internet, the commentator who was decoding and analysing all the moves live goggled in bewilderment, placed the stone on his demonstration board, then removed it hesitantly. He checked on his screen and then finally replaced it. 'It's a very surprising move!' he said with a perplexed smile. 'I thought it was a mistake,' the other commentator replied.* In the four corners of the world, the game's greatest specialists were expressing the same amazement. Had the computer just made a huge mistake, or was it on the other hand just a stroke of genius? Three and a half hours and one hundred and seventy-four moves later, the answer was incontrovertibly given. The South Korean champion resigned – the machine had won.

After the match, the adjectives used to describe move 37 came pouring out: 'creative', 'unique', 'fascinating'.

No human being would have made such a move, which the traditional strategies considered to be a poor one but had just led to victory. The question then arises: how can a computer, which merely follows an algorithm written by human beings, exhibit evidence of creativity?

The answer to this question lies in new types of training algorithms. The programmers did not in fact teach the computer how to play. They taught it how to learn to play! During its training sessions, AlphaGo spent thousands of hours playing against itself and detecting the moves that led to victory on its own. The computer also relied on the fact

---

* A recording of this moment can be viewed online on https://www.youtube.com/watch?v=l-GsfyVCBu0&feature=youtu.be&t=1h17m50s

that chance was introduced into its algorithm. There are far too many possible combinations in Go for them all to be calculable, even by a computer. To remedy this, AlphaGo selects the paths it will explore at random and uses probability theory. The computer tests only a small sample of all the possible combinations and determines the moves that have the greatest chance of victory in the same way that a poll estimates the characteristics of an entire population from a small group. This is part of the secret behind AlphaGo's intuition and originality: not to think systematically, but to weigh up the possible outcomes according to their probabilities.

Over and above games of strategy, computers, equipped with increasingly complicated and efficient algorithms, now appear able to outperform people in most of their skills. They drive cars, take part in surgical operations, and are able to compose music or paint original pictures. It is hard to imagine a human activity that, from a technical point of view, cannot be performed by a machine piloted by an adaptive algorithm.

Given these dazzling advances accomplished in just a few decades, who knows what the computers of the future will be capable of? And who knows whether one day they will be able to invent new mathematics on their own? For the moment, the game of mathematics remains too complicated for computers to give free rein to their creativity in it. Their usage in mathematics continues to be for technical purposes and calculation. But perhaps one day a descendant of AlphaGo will produce a remarkable theorem which, like its ancestor's move 37, will leave all the great scholars on the planet dumbstruck. It is difficult to predict what the achievements of tomorrow's machines will be, but it would be surprising if they did not surprise us.

17

MATHS TO COME

The sky is dark and the sound of the rain reverberates over the roofs of Zurich. What a dismal day it is in midsummer! The train shouldn't be long now.

It is Sunday, 8 August 1897, and a man with a pensive air is standing on the station platform waiting for his guests. Adolf Hurwitz is a mathematician. Of German origin, he has now been settled in Zurich for five years, where he occupies the chair of mathematics at the Eidgenössische Technische Hochschule (ETH). It is under this title that he has played an important role in organizing the event that is to be held over the next three days. The train that is due will set down on this platform a sample of the greatest scholars of the world, coming from sixteen different countries. Tomorrow, the very first International Congress of Mathematicians will open.

The two initiators of this congress were the Germans Georg Cantor and Felix Klein. The former became famous for discovering that there exist infinities that are larger than others, and for developing set theory, so that sets could be handled without the user falling into paradoxes. The second was a specialist in algebraic structures. Even though, for diplomatic reasons, Switzerland was selected as host country for this first congress, it was not surprising that the initiative had come from Germany. In the nineteenth century, the country had made its name as the new El Dorado of mathematics. Göttingen and its prestigious university were the nerve centre where the brilliant minds of the discipline gathered.

The two hundred participants at the congress also included a good number of Italians such as Giuseppe Peano, who was known for having defined the modern axioms of arithmetic, Russians such as Andrey Markov, whose works revolutionized the study of probability, and French people such as Henri Poincaré,* who discovered, amongst other things, what came to be known as chaos theory and what was later called the butterfly effect. For the three days of the Congress, all these high-flying people had the opportunity to enter into discussions, exchange views and create links among themselves and between their domains of research.

At that time, at the end of the nineteenth century, the mathematical world was going through a transformation. With the expansion of the discipline, both geographically and intellectually, scholars had grown more remote from one another. Mathematics was becoming too vast for a single individual to be able to embrace its whole extent. Henri Poincaré, who gave the opening speech at the congress, is sometimes considered to have been the last great universal scholar, for his mastery of all the mathematics of his period and for having produced significant

---

* We have already met Poincaré. It is to him that we owe the phrase: 'Mathematics is the art of giving the same name to different things.'

advances in a variety of different areas. The species of generalists was to die out with him, to be replaced by that of specialists.

However, as though in reaction to this inexorable drifting apart of the mathematical continents, the researchers worked harder than ever to increase the number of opportunities for working together and making their discipline a united and indivisible block. Thus, mathematics entered the twentieth century torn between these two contradictory impulses.

The second International Congress of Mathematicians took place in Paris in August 1900. Subsequently, the event became established in the calendar with a congress every fourth year, apart from several exceptions caused by cancellations due to the world wars. The most recent congress was held in Seoul from 13 to 21 August 2014. With more than five thousand participants attending from one hundred and twenty different countries, that meeting was the largest gathering of mathematicians ever organized. At the time of writing, the next one is due to take place in Rio de Janeiro in August 2018.

Over the years, certain traditions have become established. For example, since 1936 the prestigious Fields medal has been awarded there. This reward, often called the Nobel Prize for mathematics, is the highest honour in the discipline. The medal itself represents a portrait of Archimedes accompanied by a somewhat high-flown quotation from the Greek mathematician: *Transire suum pectus mundoque potiri* ('To rise above oneself and grasp the world').

Profile of Archimedes on the Fields medal.

Another effect of this mathematical globalization is that English has gradually established itself as the international language of the discipline. It is worth noting that as early as the Paris Congress, some participants had complained that giving the conferences and their abstracts exclusively in French was an obstacle to their comprehension by foreign participants. The Second World War and the exodus of large numbers of European scholars to the United States and its major universities played a large part in this movement. Nowadays, the overwhelming majority of mathematical research articles are written and published in English.*

The number of mathematicians has also risen considerably in the last century. In 1900, there were only a few hundred, chiefly in Europe. Today, there are tens of thousands in the four corners of the world. Several dozen new articles are published each day. According to some estimates the worldwide mathematical community currently produces around a million new theorems every four years!

The unification of mathematics also involved an extensive reorganization of the discipline itself. One of the most active architects of this movement was the German David Hilbert. A professor at the University of Göttingen, Hilbert was, along with Poincaré, one of the most brilliant and most influential mathematicians of the early twentieth century.

In 1900, Hilbert attended the Paris Congress, where on Wednesday, 8 August he gave a speech at the Sorbonne that has now become famous. There, the German mathematician presented a list of significant

---

* Since 1991, these articles from all over the world have been freely available on the Internet via the platform arXiv.org which was put in place by Cornell University in the United States. If you would like to see what a mathematics article looks like, go there and take a look.

unresolved problems which, he believed, should give direction to the mathematics of the new century. Mathematicians like challenges and the initiative took off. Hilbert's twenty-three problems provoked and stimulated the interest of researchers and very soon spread far beyond those who attended the conference.

In 2016, four of the problems still remained unanswered. Of these, the eighth problem on Hilbert's list, known as the Riemann hypothesis, is generally considered to be greatest mathematical conjecture of our age. It involves finding the imaginary solutions to an equation set in the mid-nineteenth century by the German Bernhard Riemann. If this equation is particularly interesting, it is because it holds the key to a much older mystery: that of the sequence of prime numbers that has been studied since antiquity.* Eratosthenes had been one of the first to study this sequence in the third century BC. Find the solutions to Riemann's equation and you will thereby obtain highly substantial information about these numbers which occupy a central place in arithmetic.

These problems lived their own lives, and Hilbert did not stop there. In the years that followed he began to put in place a vast programme to formulate all of mathematics on a single, solid, reliable and definitive logical foundation. His objective was to create a unique theory that would cover all branches of mathematics! You will recall that after Descartes and his coordinates, problems in geometry could then be expressed in the language of algebra. Thus, geometry had in some sense become a subdiscipline of algebra. But was it possible to repro-

---

* The prime numbers are numbers that cannot be written as the product of two numbers smaller than themselves. For example, 5 is a prime number. The sequence of prime numbers begins with 2, 3, 5, 7, 11, 13, 17, 19 . . .

duce this fusion of disciplines on the scale of mathematics as a whole? In other words, was it possible to find a super-theory in which all branches of mathematics, from geometry to probability, via algebra and infinitesimal calculus, would be simply special cases?

This super-theory effectively materialized with the revisiting of the framework of set theory established at the end of the nineteenth century by Georg Cantor. Several proposals for the axiomatization of this theory were outlined at the start of the twentieth century. Between 1910 and 1913, the two Britons Alfred North Whitehead and Bertrand Russell published a work in three volumes entitled *Principia Mathematica*, in which they set down the axioms and the logical rules based on which they recreated the rest of mathematics from scratch. One of the book's most famous passages is to be found on page 362 of the first volume where, after having recreated arithmetic, Whitehead and Russell, finally arrive at the theorem 1 + 1 = 2! The commentators were highly amused by the fact that it took so many pages and so much argumentation that was incomprehensible to novices to reach such an elementary equation. For your visual delectation, here is what the proof of 1 + 1 = 2 looks like in the symbolic language of Whitehead and Russell.

$*54\cdot43.$ $\vdash:. \alpha, \beta \epsilon 1 . \supset : \alpha \cap \beta = \Lambda . \equiv . \alpha \cup \beta \epsilon 2$

Dem.

$\vdash . *54\cdot26 . \supset \vdash :. \alpha = \iota^{\prime}x . \beta = \iota^{\prime}y . \supset : \alpha \cup \beta \epsilon 2 . \equiv . x \neq y .$

$[*51\cdot231]$ $\equiv . \iota^{\prime}x \cap \iota^{\prime}y = \Lambda .$

$[*13\cdot12]$ $\equiv . \alpha \cap \beta = \Lambda$ (1)

$\vdash . (1) . *11\cdot11\cdot35 . \supset$

$\vdash :. (\exists x, y) . \alpha = \iota^{\prime}x . \beta = \iota^{\prime}y . \supset : \alpha \cup \beta \epsilon 2 . \equiv . \alpha \cap \beta = \Lambda$ (2)

$\vdash . (2) . *11\cdot54 . *52\cdot1 . \supset \vdash . \text{Prop}$

From this proposition it will follow, when arithmetical addition has been defined, that $1 + 1 = 2$.

Please don't try to understand anything at all in this agglutination of symbols; it is impossible without having read the previous 361 pages!*

After Whitehead and Russell, other proposals to improve the axioms were put forward, and today the vast bulk of modern mathematics is indeed founded upon the small number of basic axioms of set theory.

Despite the astonishing success of set theory, Hilbert was still not satisfied, as a few doubts remained about the trustworthiness of the axioms of *Principia Mathematica*. For a theory to be considered to be perfect, it must satisfy two criteria: it must be consistent and complete.

Consistency means that the theory does not admit paradoxes. It is not possible to prove both a statement and its negation in it. If, for example, one of the axioms can be used to prove that $1 + 1 = 2$ and another leads to $1 + 1 = 3$, the theory is inconsistent since it contradicts itself. Completeness, on the other hand, asserts that the axioms of the theory are sufficient to prove all that is true in the theory. If, for example, an arithmetic theory does not have sufficient axioms to prove that $2 + 2 = 4$, then it is incomplete.

Was it possible to show that *Principia Mathematica* satisfied these two criteria? Could one be certain that one would never find paradoxes in it and that its axioms were sufficiently precise and powerful that all possible and imaginable theorems could be deduced from them?

Hilbert's programme was brought to a halt in a manner that was as brutal as it was unexpected when in 1931 a young Austrian mathematician by the name of Kurt Gödel published an article entitled 'Über formal unentscheidbare Sätze der *Principia Mathematica* und verwandter Systeme', or 'On Formally Undecidable Propositions of *Principia*

---

* And even after reading them, it is frankly not easy . . .

Mathematica and Related Systems'. This article proved an extraordinary theorem which asserted that there could not exist a super-theory that was both consistent and complete! If the *Principia Mathematica* is consistent, then there necessarily exist so-called undecidable assertions that can be neither proved nor refuted based on them. It is thus impossible to determine whether they are true or false!

GÖDEL'S EXQUISITE CATASTROPHE

Gödel's incompleteness theorem was a landmark in mathematical thinking. To try to understand its general principle, we need to take a detailed look at how we write mathematics. Here are two elementary assertions in arithmetic:

A. THE ADDITION OF TWO EVEN NUMBERS ALWAYS GIVES AN EVEN NUMBER.

B. THE ADDITION OF TWO ODD NUMBERS ALWAYS GIVES AN ODD NUMBER.

These two statements are quite clear; it would not be difficult to write them in the algebraic language of Viète. When you think about it a little, you will see that the first of these assertions, denoted by A, is true, while the second, denoted by B, is false, since the sum of two odd numbers is always even. This leads us to the following two new statements:

C. ASSERTION A IS TRUE.

D. ASSERTION B IS FALSE.

These two new phrases are a little peculiar. Properly speaking, they are not mathematical statements, but statements that talk about mathematical statements. The phrases C and D, unlike A and B, cannot a priori be written in the symbolic language of Viète. Their subjects are neither numbers, nor geometric figures, nor any other object of arithmetic, probability or infinitesimal calculus. They are so-called metamathematical statements, that is to say statements that do not talk about mathematical objects, but about mathematics itself. A theorem is mathematical. The assertion that the theorem is true is metamathematical.

The distinction may seem subtle and ludicrous, yet it was through an incredibly ingenious formalization of metamathematics that Gödel obtained his theorem. His achievement was to find a way of writing metamathematical statements in the language of mathematics itself. Based on a brilliant process that made it possible to interpret statements as numbers, mathematics was suddenly empowered to talk about itself, as well as about numbers, geometry and probability.

What does something that talks about itself remind you of? Do you remember the famous paradox of Epimenides? The Greek poet one day asserted that all Cretans were liars. Since Epimenides was himself a Cretan, it was impossible to determine whether his declaration was true or false without coming up against a contradiction. One thinks of the snake that bites its own tail. Until that day, mathematical statements had been spared from self-referencing assertions of that type. But based on his process, Gödel managed to reproduce a phenomenon of the same type in the very heart of mathematics. Consider the following statement:

G. THE STATEMENT G IS NOT PROVABLE FROM THE AXIOMS OF THE THEORY.

This statement is manifestly metamathematical, but using Gödel's trick, it can nevertheless be expressed in the language of mathematics. It thus becomes possible to try to prove G from the axioms of the theory. And then, two scenarios emerge.

First scenario: G is provable. But in this case, since G asserts that it is not provable, this means that it is mistaken, thus that it is false. Now, if it is possible to prove something false, then the theory as a whole collapses! It is not consistent.

Second scenario: G is not provable, in which case, what G says is true, which means that our axioms are incapable of proving an assertion, which is indeed true! The theory is therefore incomplete since there exist true statements that are inaccessible to it.

To sum up, we are losers in all cases. Either the theory is inconsistent, or it is incomplete. Gödel's incompleteness theorem demolished Hilbert's sweet dreams definitively. And it was useless to try to bypass the problem by changing theories, because its result applied not only to *Principia Mathematica*, but also to any other theory that might claim to be able to replace it. A unique and perfect theory in which all theorems are provable cannot exist.

However there remained a hope. The statement G is certainly undecidable, but one has to admit that it is not very interesting from a mathematical point of view. It is a curiosity that Gödel created from scratch so as to be able to exploit the flaw in Epimenides' statement. However, it was still possible to hope that the great problems of mathematics, those that are interesting, might not fall into the trap of self-reference.

Unfortunately, once again it became necessary for mathematicians to resign themselves to the situation. In 1963, the American mathematician Paul Cohen proved that the first of Hilbert's twenty-three problems also belonged to the strange category of undecidable statements. It was impossible to prove it or to refute it based on the axioms of the *Principia Mathematica*. If this problem were one day to be resolved, that would necessarily take place in the framework of a different theory. But that new theory would then also contain flaws and other undecidable statements.

While studies of the foundations of mathematics occupied an important place in the twentieth century, that did not prevent other branches of the discipline from pursuing their own paths. It is difficult to describe the abundant diversity of mathematics that has developed over the recent decades. Let us, however, pause again for few moments to look at one of the most dazzling gems of the last century: the Mandelbrot set.

This amazing creature stems from analysing the properties of certain numerical sequences. Choose a number, any number you like, then construct a sequence for which the first term is 0 and in which every term is then equal to the square of the previous term plus your chosen number. If, for example, you choose the number 2, then your sequence will begin as follows: 0, 2, 6, 38, 1,446 . . ., where you will note that $2 = 0^2 + 2$, then $6 = 2^2 + 2$, then $38 = 6^2 + 2$, then $1,446 = 38^2 + 2$, and so on. If instead of 2 you were to choose -1, then you would obtain the sequence 0, -1, 0, -1, 0 . . . This sequence simply alternates between 0 and -1, since $-1 = 0^2 - 1$ and $0 = (-1)^2 - 1$.

These two examples show that the sequence obtained may exhibit two very different behaviours depending on the number chosen. The sequence may take larger and larger values and fly off to infinity, as happens if you take the number 2. On the other hand, the sequence may be bounded (that is to say, its values do not move away and remain within a limited region), as is the case for the number −1. All numbers, be they integers, decimal numbers or even imaginary numbers, can thus be assigned to one category or the other.

This classification of numbers may seem quite abstract, so in order to visualize things better, it can be represented geometrically using Cartesian coordinates. In the plane, we place all the real numbers on a horizontal axis as we already did before,* then the imaginary numbers on a vertical axis. We can now colour the points belonging to the two categories with different colours. At this stage, a marvellous figure appears.

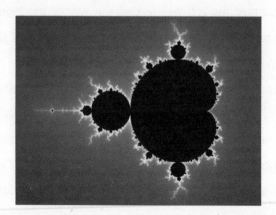

In this figure, the numbers coloured in black are those that generate bounded sequences, while those in grey are those that yield sequences that go to infinity. A white 'shadow' has been inserted behind the black

---

* Zero in the middle, the negative numbers to the left and the positive numbers to the right.

figure so that certain extremely fine details that are often invisible to the naked eye can be detected more easily.

Each point of the image corresponds to the calculation and study of a sequence, and the drawing of this figure requires a very large number of calculations. That is why it only became possible to obtain precise representations at the start of the 1980s using computers. The French mathematician Benoît Mandelbrot was one of the first to study in detail the geometry of this figure, which his colleagues ultimately named after him.

The Mandelbrot set is fascinating! Its contours constitute an unlikely piece of geometric lacework, full of harmony and precision. If you zoom in on its edges, you will see more and more infinitely fine and incredibly delicate patterns. In truth, it is almost impossible for a single image to capture the full richness of the shapes the Mandelbrot set reveals when you dissect it in detail. A small sample of these details can be seen on the figure on the next page.

But what makes it even more remarkable is the disarming simplicity of its definition. If – in order to draw this figure – we had needed to call upon monstrous equations, erudite and chaotic calculations or bizarre constructions, we might have been able to say: 'Certainly, the figure is beautiful, but it's completely artificial and of little interest.' But no, this figure is simply the geometric representation of elementary properties of numerical sequences that are defined in just a few words. This geometric marvel issues from a very simple rule.

A discovery of this type inevitably reopens the debate about the nature of mathematics: is it a human invention or does it have an independent existence? Do mathematicians discover it or create it? At first sight, the Mandelbrot set appears to argue in favour of discovery.

If this figure takes this extraordinary form, it is not because Mandelbrot decided to construct it in that way. He had at no point wanted to invent such a figure. It just imposed itself on him. It could not have been anything other than what it is.

However, it is still a very strange thing to consider the existence of an object that is not only purely abstract, but whose interest does not go beyond the intangible framework of mathematics. While numbers, triangles and equations are abstract, they can help us to comprehend the real world. Abstraction until now has always appeared to have retained a reflection, however distant, in the tangible universe. The

Mandelbrot set no longer appears to have any direct link to that. No known physical phenomenon adopts a structure that even remotely resembles it. Then why is it of interest? Can its discovery be placed on the same level as the discovery of a new planet in astronomy, or a new animal species in biology? Is it an object worth studying for its own sake? In other words, does mathematics work on a level playing field with the other sciences?

Many mathematicians will undoubtedly reply 'Yes' to this question. However, the discipline occupies an extremely special position in the field of human knowledge. One of the reasons for this specialness lies in the ambiguous relationship between mathematics and the beauty of its objects.

It is true that particularly beautiful things are to be found in almost all the sciences. The images of heavenly bodies we receive from astronomers are an example of this. We marvel at the shape of galaxies, the brilliant tails of comets and the vivid colours of nebulae. The Universe is beautiful, certainly. That is by chance. But it has to be said that if it had not been, we could not have done much about it. Astronomers don't have a choice. The heavenly bodies are what they are, and it would have been necessary to study them even if they had been ugly. Of course, the definitions of beauty and ugliness are very subjective, but that is not the issue here.

Mathematicians, on the other hand, seem a little freer. As we have already seen, there are infinitely many ways of defining algebraic structures. And in each of these there are infinitely many ways of defining sequences whose properties can be studied. Most of these routes will not lead to sets that are as beautiful as the Mandelbrot set. In mathematics, we have a much greater freedom to choose what we study. Among the infinitely many theories that could be explored, it is very often those that appear the most elegant that we choose.

This approach appears to resemble an artistic one. If Mozart's symphonies are beautiful, it is not by chance, but because the Austrian composer did everything he could to make them so. Of the infinitely many pieces of music that could be composed, the vast majority are ugly. Depress the keys of a piano at random and you will be convinced of that. The artist's talent lies in his or her ability to find the few gems that will amaze us among the infinitely many possibilities that are of no interest.

In the same way, it is a part of the mathematician's talent to be able to find the objects most worthy of interest in the infinitude that is the mathematical world. It is evident that if the Mandelbrot figure had not been so beautiful, mathematicians would have been much less interested in it. It would have remained anonymous among all the other neglected figures, like all those poor symphonies that no one will ever play.

So are mathematicians artists rather than scientists? It would be going a tad too far to say that. But does the question make sense? Scientists search for truth, and sometimes find beauty there by chance. Artists search for beauty, and sometimes find truth there by chance. Mathematicians, for their part, have moments when they seem to forget that there is a difference between the two. They search for both at once. Whichever they find leaves them unconcerned. They mix the true and the beautiful, the useful and the superfluous, the ordinary and the unlikely, like so many colours blending on the infinite canvas.

Much like the craftsmen of Mesopotamia, modern mathematicians do not always fully fathom what they are doing. Mathematics often does not reveal its secrets or true nature until long after its creators have vanished. When Pythagoras, Brahmagupta, al-Khwārizmī, Tartaglia, Viète and all the others invented new mathematics, they could not have imagined everything that would be done with it as of today. Perhaps we are also unable to imagine everything that it will be possible to do with it in the centuries to come. Only time can provide the hindsight necessary to appreciate the full value of the corpus of mathematics.

EPILOGUE

Our account is nearing its end.

At least, it is nearing the end of the portion that I can recount now as I write this book at the beginning of the twenty-first century. What will happen after that? It is clear that the story is not over.

This is something to accept when you do science: that the more you know about a subject, the more you measure the reach of our ignorance. Every answer given raises ten new questions. This endless game is both overwhelming and exhilarating. It must be said: if it were possible for us to know everything, the delight that would result from this would be eclipsed at once by the much greater sadness of having nothing left to discover. But no need for alarm. Thankfully, the mathematics that remains to be done is undoubtedly vaster than what we already know.

What will the mathematics of the future look like? This question makes one's head spin. It is dizzying to walk towards the very edge of our knowledge and to look out over the expanse of everything that we do not know! For anyone who has tasted even just once the intoxicating flavour of new discoveries, the call of unknown lands is undoubtedly stronger than the comfort of conquered territories. Mathematics is so fascinating when it is still untamed! And what a delight it is to observe, in the distant haze, wild ideas bounding freely around in the infinite savannah of our ignorance. There is something sublime about them, and their mystery torments our imagination deliciously. Some of them seem close. We could imagine that we have only to stretch out a hand to be able to touch them. Others are so far away that it will take generations to get near them. No one knows what the mathematicians of the centuries to come will discover, but it is worth betting it will be full of surprises.

It is now May 2016, and I am walking through the alleyways of the Salon Culture et Jeux Mathématiques, an exhibition with numerous activities, games and competitions on the subject of mathematics and its history, which takes place every year in the 6th arrondissement in Paris. This is a place I particularly like. There are magicians here who explain card tricks to you based on an arithmetic property. There are sculptors working on stone sculptures with geometric structures inspired by the Platonic solids. There are also inventors whose wooden mechanisms form strange calculating machines. A little further on, I come across a few people calculating the radius of the Earth by reproducing the experiment of Eratosthenes. I then notice a stand for fans of origami, one for lovers of brainteasers and one for calligraphers. A play involving maths and astronomy is being performed in the marquee, from where great bursts of laughter are escaping.

All these people are doing mathematics. All these people are inventing mathematics, each in their own way! This juggler is going to use geometric figures for his performance, which no great scientist would have deemed worthy of interest. But for him, they are beautiful, and his spheres, which race around in the air, bring a sparkle to the eyes of passers-by.

I think that all that is even more heartening than all the great discoveries of great scholars. Mathematics, even at its simplest, provides an inexhaustible source of astonishment and wonder. The visitors to the Salon include many parents who are attending primarily for their children's sake, but who gradually get caught up themselves. It is never too late. Mathematics has a formidable potential for becoming a discipline that is popular and fun. You don't have to be a mathematical genius to be passionate about it and enjoy the exploration and the discoveries.

You don't need very much in order to do mathematics. And if you feel like continuing after you have turned over the page, you will discover a lot more than everything I have been able to tell you about. You will be able to map out your own path, establish your own tastes and follow your own desires.

And all you will need is a hint of audacity, a good dash of curiosity, and a little imagination.

TO GO FURTHER

To extend your exploration of mathematics, here are a few leads that you may find useful.

MUSEUMS AND EVENTS

There are several maths-themed institutions around the world. New York City has the **MoMaths** museum (http://momath.org) while Germany is host to the **Mathematikum** (http://www.mathematikum.de), in Gießen – both venues are exclusively devoted to mathematics.

In the United Kingdom, there is no museum or gallery purely dedicated to the subject (something which an organisation called 'MathsWorldUK' is currently trying to redress by establishing the 'first national mathematics discovery centre'). However, there are sections of museums that can be visited. For instance, the **Science Museum** also has a special section, the Winton Gallery (launched in 2016), which was designed by the late architect Zaha Hadid, who incidentally was a trained mathematician! The Winton Gallery's mission is to showcase objects that 'reveal how mathematics connects to every aspect of our lives from war and peace to life, death, money, trade and beauty'. Otherwise, the **British Museum** contains artefacts such as abacuses, vases and coins, which can be browsed online or viewed in person. **The V&A Museum of Childhood** is home to an abacus from the 1920s, on which children and elders alike can (re)discover a fundamental way of counting.

In terms of events, (although there are some dissenting voices in the UK, due to the fact that π converted into a date assumes an American reading of the calendar) there are a lot of celebrations for π day on 14 March, worldwide. This is probably the largest global celebration of maths. In France, the 'Semaine des mathématiques' in March is a weeklong celebration with additional fringe events. 'The Fête de la Science' (http://www.fetedelascience.fr), is also held in October. Other one-off events include the 'Salon Culture & Jeux Mathématiques' (www.cijm.org), which takes place annually at the end of May in Paris. The mathematics department of the Palais de la Découverte in Paris (http://www. palais-decouverte.fr) also offers activities, talks and workshops for the general public. If you are passing by, don't miss looking around the famous π Room! Again in Paris, the Cité des sciences et de l'industrie (http://www. cite-sciences. fr) also has a space dedicated to mathematics.

Meanwhile, England has the Cheltenham Science Festival with STEM activities and talks, which are primarily for adults. For the younger crowd, there's the MathsFest (for sixth formers) and the Oxford Maths Festival, both of which have inclusive family-friendly activities to instil a love of maths from a young age, or reignite it; 2018 activities include crafty maths, puzzles and board games. It is also worth keeping an eye out on the London Mathematical Society's events page, for more in-depth talks: (https://www.lms.ac.uk/content/ calendar).

BOOKS

There are large numbers of works dealing with mathematics at various levels from the popular to the specialized. The following suggestions are not exhaustive but are ones that stood out while writing this book.

Martin Gardner, who was Mathematical Games columnist for *Scientific American* from 1956 to 1981, was a key figure in recreational mathematics. The collections of his columns, together with his numerous books on mathematical magic or enigmas, are reference works in the field. Among the classics, one might also mention Yakov Perelman and his famous *Mathematics Can Be Fun*, and Raymond Smullyan, with his books on logic such as *The Lady or the Tiger* and *What is the Name of This Book?*

Among more recent authors, we recommend the books of Ian Stewart, such as *Professor Stewart's Cabinet of Mathematical Curiosities*, Marcus Du Sautoy, with among others *Symmetry: A Mathematical Journey*, or Simon Singh's *The Code Book* or *The Simpsons and their Mathematical Secrets*. *The Math Book* by Clifford A. Pickover offers an illustrated chronological panorama of the most fabulous gems of the history of mathematics.

As far as French authors are concerned, we would mention in particular Denis Guedj, whose many works include the famous historico-mathematical detective novel *The Parrot's Theorem* (in translation). Jean-Paul Delahaye, whose books include *Le Fascinant Nombre π* and *Merveilleux nombres premiers,* is another inspiring author.

In another genre, *Birth of a Theorem, a Mathematical Adventure* (in translation), by Cédric Villani, provides an insight into modern mathematical research.

ONLINE

The website 'Image des mathématiques' (http://images.math.cnrs.fr) offers regular articles (in French) by mathematicians popularizing current mathematical research.

The films *Dimensions* (available in English through http://www.dimensions-math.org) and *Chaos* (available in English through http://www.chaos-math.org), produced by Jos Leys, Aurélien Alvarez and Étienne Ghys, take you into the world of the fourth dimension and chaos theory with magnificent animations.

For a number of years, popular science channels have been growing in number, particularly on YouTube. In mathematics, I would mention the YouTube channel 'Numberphile' and the videos of Vi Hart. It's also worth checking out 'Science4All' (in English, http://www.science4all.org/), 'Statistics explained to my cat' (in English) and 'Passe-Science' (in French).

You could also search for videos of public talks given by researchers in mathematics. The mathematicians Étienne Ghys, Tadashi Tokieda and Cédric Villani are particularly brilliant presenters.

BIBLIOGRAPHY

Here is a list of the main documents referred to. Note that some may be very technical. The list is given in the alphabetical order of the authors.

KEY:	
ERA	**SUBJECT:**
A : ANTIQUITY	G : GEOMETRY
M : MIDDLE AGES	N : NUMBERS/ALGEBRA
R : RENAISSANCE	P : ANALYSIS/PROBABILITY
E : MODERN ERA & CONTEMPORARY	L : LOGIC
	S : OTHER SCIENCES

[**EP**] M. G. Agnesi, *Traités élémentaires de calcul différentiel et de calcul intégral*, Claude-Antoine Jombert Libraire, 1775 (French translation from original Italian)

D. J. Albers, G. L. Alexanderson and C. Reid, *International Mathematical Congresses, an illustrated history*, Springer-Verlag, 1987

[**AG**] Archimedes, *Works of Archimedes*, translated and edited by T. L. Heath (1897), Dover Publications, 2002

[**AL**] Aristotle, *Physics*, translated by Robin Waterfield, Oxford World's Classics, 2008

[**EP**] S. Banach and A. Tarski, 'Sur la décomposition des ensembles de points en parties respectivement congruentes', *Fundamenta Mathematicae*, 1924 (in French)

[**E**] B. Belhoste, *Paris savant*, Armand Colin, 2011 (in French)

[**EP**] J. Bernoulli, *The Art of Conjecturing*, translated by Edith Dudley Sylla, Johns Hopkins University Press, 2006

[**G**] J.-L. Brahem, *Histoires de géomètres et de géométrie*, Éditions le Pommier, 2011 (in French)

[**MN**] H. Bravo-Alfaro, 'Les Mayas, un lien fort entre Maths et astronomie', in Aurélien Alvarez, Rémi Anicotte, Jean-Marc Bonnet-Bidaud, Sonja Brentjes, H. Bravo-Alfaro, et al., *Maths Express au carrefour des cultures*, Marc Moyon, France, CIJM – Comité International des Jeux Mathématiques, 2014 (in French)

[**N**] F. Cajori, *A History of Mathematical Notations*, The Open Court Company, 1928

[**RN**] G. Cardano, *The Rules of Algebra (Ars Magna)*, Dover Publications, 1968

[**RN**] L. Charbonneau, 'Il y a 400 ans mourait sieur François Viète, seigneur de la Bigotière', *Bulletin AMQ*, 2003 (in French)

[**AG**] K. Chemla, G. Shuchun, *Les Neuf Chapitres, le classique mathématique de la Chine ancienne et ses commentaires*, Dunod, 2005 (in French)

[**AG**] K. Chemla, 'Mathématiques et culture, Une approche appuyée sur les sources chinoises les plus anciennes connues', *La mathématique, 1, Les lieux et les temps*, CNRS Éditions, 2009 (in French)

[**AG**] M. Clagett, *Ancient Egyptian Science: A Source Book*, American Philosophical Society, 1999

[**EG**] R. Cluzel and J.-P. Robert, *Géométrie – Enseignement technique*, Librairie Delagrave, 1964 (in French)

Collective – Department of Mathematics – North Dakota State University, Mathematics Genealogy Project, https://genealogy.math.ndsu.nodak.edu/, 2016

[**N**] J. H. Conway and R. K. Guy, *The Book of Numbers*, Springer-Verlag, 1996

[**E**] G. P. Curbera, *Mathematicians of the World, Unite!: The International Congress of Mathematicians – A Human Endeavor*, CRC Press, 2009

J.-P. Delahaye, *Le fascinant nombre π*, Belin, 2001 (in French)

A. Deledicq et al., *La longue histoire des nombres*, ACL – Les éditions du Kangourou, 2009 (in French)

[**AG**] A. Deledicq and F. Casiro, *Pythagore & Thalès*, ACL – Les éditions du Kangourou. 2009 (in French)

A. Deledicq, J.-C. Deledicq and F. Casiro, *Les maths & la plume*, ACL – Les éditions du Kangourou, 1996 (in French)

[**M**] A. Djebbar, 'Bagdad, un foyer au carrefour des cultures', in Alvarez et al., as cited above under Bravo-Alfaro (in French)

[M] A. Djebbar, 'Les mathématiques arabes', in *L'âge d'or des sciences arabes* (collectif), Actes Sud – Institut du Monde Arabe, 2005 (in French)

[M] A. Djebbar, 'Panorama des mathématiques arabes', in *La mathématique, les lieux et les temps*, CNRS Éditions, 2009 (in French)

[A] D. W. Engels, *Alexander the Great and the Logistics of the Macedonian Army*, University of California Press, 1992

[EG] Euclid, *The Thirteen Books of Euclid's Elements*, translated by T. L. Heath (1908), 3 vols, Dover, 1956

[EN] P. Eymard and J.-P. Lafon, *The Number π*, translated by Stephen S. Wilson, AMS, 2004

[MN] L. Fibonacci, *Fibonacci's Liber Abaci*, translated by Laurence Sigler, Springer-Verlag, 2010

[ES] Galileo, *The Assayer*, English translation by S. Drake, http://www.princeton.edu/~hos/h291/assayer.htm

[MG] R. P. Gomez et al., *La Alhambra*, Epsilon, 1987 (in French)

[N] D. Guedj, *Zéro*, Pocket, 2008 (in French)

B. Hauchecorne, *Les Mots & les Maths*, Ellipses, 2003 (in French)

B. Hauchecorne and D. Surreau, *Des mathématiciens de A à Z*, Ellipses, 1996 (in French)

[E] D. Hilbert, 'Mathematical Problems', *Bull. Amer. Math. Soc.* 8 (1902), 437–79

[EL] D. Hofstadter, *Gödel, Escher, Bach: An Eternal Golden Braid*, Basic Books, 1999

[AN] J. Høyrup, *Length, Widths, Surfaces*, Springer-Verlag, 2013

[AN] J. Høyrup, *L'algèbre au temps de Babylone*, Vuibert/ADAPT–SNES, 2010 (in French)

[AN] J. Høyrup, 'Les Origines', in *La mathématique, les lieux et les temps*, CNRS Éditions, 2009 (in French)

[A] Jamblique, *Vie de Pythagore, La roue à livres*, 2011 (in French)

[N] M. Keith after E. A. Poe, 'Near a Raven', www.cadaeic.net/naraven.htm, 1995

[MN] A. Keller, 'Des devinettes mathématiques en Inde du Sud', in Alvarez et al., cited above under Bravo-Alfaro (in French)

[MN] M. al-Khwārizmī, *Algebra*, English translation by Frederic Rosen, Oriental Translation Fund, 1831

[A] D. Laërce, *Vie, doctrines et sentences des philosophes illustres*, GF-Flammarion, 1965 (in French)

[EP] M. Launay, *Urnes Interagissantes*, Doctoral thesis, Aix-Marseille University, 2012 (in French)

[**EG**] B. Mandelbrot, *Les objets fractals*, Champs Science, 2010 (in French)

[**E**] J.-C. Martzloff, *A History of Chinese Mathematics*, translated by Stephen S. Wilson, Springer-Verlag (1997)

S. Mehl, ChronoMath, chronologie et dictionnaire des mathématiques, http://serge.mehl.free.fr/ (in French)

[**M**] M. Moyon, 'Traduire les mathématiques en Andalus au XIIe siècle', in Alvarez et al., cited above under Bravo-Alfaro.(in French)

[**EL**] E. Nagel and J. R. Newman, K., *Gödel's Proof*, NYU Press, 2001

[**RN**] P. D. Napolitani, 'La Renaissance italienne', in *La mathématique, les lieux et les temps*, CNRS Éditions, 2009 (in French)

[**ES**] I. Newton, *Principia: The Mathematical Principles of Natural Philosophy* (1687), reprinted Prometheus Classics, 2017

[**EP**] B. Pascal, *Traité du triangle arithmétique*, Guillaume Desprez, 1665 (in French)

A. Peters, *Histoire mondiale synchronoptique*, Éditions académiques de Suisse – Bâle 1966 (in French)

[**AG**] Plato, *Timaeus*, translated by Robin Waterfield, Oxford World's Classics, 2008

[**MN**] K. Plofker, *Mathematics in India*, Princeton University Press, 2009

[**E**] H. Poincaré, *Science and Method*, English translation by Francis Maitland, Cosimo Classics, 2007

[**EP**] G. Pólya, 'Sur quelques points de la théorie des probabilités', *Annales de l'Institut Henri Poincaré*, 1930 (in French)

[**AN**] C. Proust, Brève chronologie de l'histoire des mathématiques en Mésopotamie, CultureMATH, http:// culturemath.ens.fr/content/brève-chronologie-de- lhistoire-des-mathématiques-en-mésopotamie, 2006 (in French)

[**AN**] C. Proust, Le calcul sexagésimal en Mésopotamie, CultureMATH, http://culturemath.ens.fr/content/ le-calcul-sexagésimal-en-mésopotamie, 2005(in French)

[**AN**] C. Proust, Mathématiques en Mésopotamie, images des mathématiques, http://images.math.cnrs.fr/ Mathematiques-en-Mesopotamie.html, 2014 (in French)

[**A**] Pythagoras, *The Golden Verses of Pythagoras*, translated by N. L.Redfield from French translation by Fabre d'Olivet, Createspace Independent Publishing Platform, 2015

[**EL**] B. Russell and A. N. Whitehead, *Principia Mathematica*, reprint Merchant Books, 2009

M. du Sautoy, *Symmetry: A Journey into the Patterns of Nature*, Harper Perennial, 2009

[**AN**] D. Schmandt-Besserat, 'From Accounting to Written Language', in B. A. Rafoth and D. L. Rubin (eds), *The Social Construction of Written Communication*, Ablex Publishing Co, Norwood, 1988

[**AN**] D. Schmandt-Besserat, The Evolution of Writing, author's personal website. https://sites.utexas.edu/dsb/, 2014.

[**RN**] M. Serfati, 'Le Secret et la Règle', in *La recherche de la vérité* (collective), ACL – Les éditions du Kangourou, 1999 (in French)

[**AG**] Shen Kangshen, *The Nine Chapters on the Mathematical Art: Companion and Commentary*, OUP (1999)

[**EL**] R. Smullyan, *Gödel's Incompleteness Theorems*, Oxford, 1982

[**EL**] R. Smullyan, *What is the Name of This Book?*, reprinted Dover, 2011

[**N**] Stendhal, *The Life of Henry Brulard*, translated by John Sturrock, NYRB Classics, 2007

[**EL**] A. Turing, 'On computable numbers with an application to the entscheidungsproblem', *Proceedings of the London Mathematical Society*, 1936

[**RN**] F. Viète, *The Analytic Art*, English translation by T. Richard Witmer, Kent State U. Press, 1983 (French original 1630)

INDEX